Cascading Logic

A Machine Control Methodology for Programmable Controllers

Cascading Logic

A Machine Control Methodology for Programmable Controllers

Gary Kirckof, P.E.

ISA–The Instrumentation, Systems, and Automation Society

Notice

The information presented in this publication is for the general education of the reader. Because neither the author nor the publisher have any control over the use of the information by the reader, both the author and the publisher disclaim any and all liability of any kind arising out of such use. The reader is expected to exercise sound professional judgment in using any of the information presented in a particular application.

Additionally, neither the author nor the publisher have investigated or considered the affect of any patents on the ability of the reader to use any of the information in a particular application. The reader is responsible for reviewing any possible patents that may affect any particular use of the information presented.

Any references to commercial products in the work are cited as examples only. Neither the author nor the publisher endorse any referenced commercial product. Any trademarks or tradenames referenced belong to the respective owner of the mark or name. Neither the author nor the publisher make any representation regarding the availability of any referenced commercial product at any time. The manufacturer's instructions on use of any commercial product must be followed at all times, even if in conflict with the information in this publication.

Printed in the United States of America.
10 9 8 7 6 5 4 3 2

ISBN 1-55617-814-X

ISA
67 Alexander Drive
P.O. Box 12277
Research Triangle Park, NC 27709

Library of Congress Cataloging-in-Publication Data
Kirckof, Gary.
Cascading logic :a machine control methodology for programmable logic controllers / by Gary Kirckof.
p. cm.
Includes bibliographical references and index.
ISBN 1-55617-814-X
1. Programmable controllers. I. Title.
TJ223.P76 K56 2002
629.8'9--dc21

2002012741

Contents

About the Author

Gary Kirckof has spent the bulk of his twenty-year career programming automated equipment and assembly lines. Cascading Logic is a direct result of that experience. He is a mechanical engineering graduate of the University of Minnesota and as such started his career engineering process piping, heating, ventilating and air conditioning systems. The shift in focus to programming happened in stages. The first step occurred after the energy crisis when energy management system became popular. These systems were designed to reduce the operating cost of the equipment he was working with so programming them became part of his job. The next step happened while installing an energy management system in a Ford Motor Company plant. While working up in the trusses he looked down at all the automated equipment below him and decided right then and there to become a machine programmer. His affinity towards programming and his mechanical engineering background made him a natural control engineer. Gary Kirckof is currently a system engineer for Remmele Engineering in St. Paul, Minnesota and is both a registered mechanical and electrical engineer.

Introduction

Cascading logic is a programming methodology for generating machine-control programs. It is intended for industrial equipment that is sequential and controlled by a programmable logic controller. The intent of Cascading Logic is to create robust and reliable programs that are easy to write, easy to debug, and easy to maintain. The goals of Cascading Logic are to save time for both the programmer and the end user of the equipment (minimize downtime) and to produce better machinery. Let's look at what industrial equipment consists of, what is needed to commission and maintain industrial equipment, and what is required of a machine-control program to best understand the reasoning behind Cascading Logic.

Typical Industrial Equipment

Industrial equipment is a collection of motions, where a *motion* is any combination of mechanisms and a prime mover that can cause a physical event to occur on the machine. Most physical events result in something actually moving, hence the name "motion" to generally describe these events; but physical events also include activities such as maintaining temperatures and communicating with databases.

Prime movers cause events to happen and include devices such as hydraulic and pneumatic cylinders, various types of motors, and heaters. Prime movers are controlled by output signals from the programmable logic controller. The output signals are coupled to the prime movers through solenoid valves, variable-speed drives, servo controllers, and so forth. The programmable logic controller receives feedback from the prime movers, mechanisms for positional information, and so forth to complete the control loop. Sensors such as proximity switches, photoelectric eyes, and thermocouples are typically used to provide the feedback.

A common machine motion is a pneumatic cylinder controlled through a solenoid valve. It uses two proximity switches to provide feedback. The programmable logic controller can command the cylinder to extend and retract, knows when the cylinder is either extended or retracted, and even knows when the cylinder is in the process of extending or retracting.

Industrial equipment is also a collection of processes. Typical processes include press-fitting two parts together, welding, bending, and part inspecting. A large amount of machine activity involves transferring parts between processes and includes pick-and-place mechanisms, dials, and walking beams. Each process on a machine and the associated part-transfer mechanisms usually require a separate machine section, referred to as a *station*. Each station is autonomous, sharing only part-transfer mechanisms and data with the other stations. As such, we can group the machine motions into processes or stations and think of the machine as a collection of stations. Its the divide-and-conquer approach that allows us to think of the machine as a whole, just a single station of the machine, or just an individual motion of a station.

The subdivision of the machine-control program follows the grouping of the machine's processes. The program will thus have a system section for the logic common to all the stations, a section for each station, plus a few miscellaneous sections to round off the code.

Typical industrial equipment is sequential, that is, the prime movers of each station are programmed to operate one after another in a prescribed manner until the required process is complete. The sequence of events are also cyclic, such that the mechanisms end up in the same positions they started from; this means the sequence can be repeated over and over again. It thus becomes the task of the programmer to convey the required machine actions and the timing of each action into a machine-control program. Programming begins not by writing logic but instead by listing each step a station must perform. A step of a sequence is the smallest action the machine controller can initiate, such as energizing a solenoid valve or triggering a vision system to start its inspection routine. The steps are then combined into a sequence for the station. This is perhaps one of the easiest programming tasks, provided you know the steps the station is required to take. It is equally important for the machine controller to have sufficient feedback, so it can execute the steps at the correct time, determine if the step finished or failed, and make other appropriate decisions.

Commissioning and Maintaining Equipment

At some point in the equipment's life, it is turned on for the first time. At this point nothing is adjusted. Devices such as flow-control valves, motor speeds, time delays inside the machine-control logic, and the position of feedback sensors all have to be adjusted and set correctly for the machine to run as required. Pneumatic cylinders may be sticky when first used, the clutch and/or brakes may need to be burnished for proper operation, and load cells will need to be calibrated. Most likely, wiring and plumbing mistakes will need to be sorted out. Everything on the machine will need to be adjusted and checked before the machine is ready for production. The programmer is also faced with time constraints to complete the equipment startup and debug as quickly as possible.

Tools can be added to the machine-control logic to help out and to speed up the commissioning process. A typical tool is a manual-mode layer added to the machine-control logic. This allows each motion to be individually and conveniently controlled from the operator-control station and provides the status of each motion, such as position. As a result, a cylinder can be repeatedly commanded to extend and retract to exercise it and to set its flow controls. A vision system can be commanded to inspect a part over and over to check its operation without the need to run the entire machine. The manual-mode layer is a great way to check out the input/output of the programmable logic controller. Each motion can be commanded to see if it goes in the correct direction and if the sensors are triggered correctly. The prime movers, sensors, plumbing, and wiring can all be checked at once. An alternative method involves triggering sensors with the tip of a screwdriver while watching the lights on the input modules and forcing outputs on and off while observing the associated prime movers. The alternate method, while widely used, does not check to see if signals make it into the controller (faulty processor racks and/or modules). It does not check association of inputs and outputs, and it does not provide convenient functionality for the end user.

Manual-mode operations can be used to walk product through the machine to set up alignment, such as hard stops, and it can be used to fine-tune sensor locations after each motion is individually checked out and initially tuned. The next step is to check out and test the logic. The logic is easier to check out when we know the motions are in working order because it is faster to determine if a problem is hardware or software. Manual-mode operations give us not only the ability to work with each motion, but provide us with a firm foundation, after we have confidence with the motions, to proceed with the logic. More tools can be incorporated into the logic to assist with program check-out.

Just as the individual motions on a machine are grouped into station sequences then into an entire machine program, the equipment can be commissioned in a like manner. Additional tools can be used to step and single cycle each station. Stepping allows an operator to walk through a station's sequence one step at a time to check the programming of the sequence and to check the alignment of the mechanisms. Single cycling allows the station to step through its sequence at rated speed to check cycle time and machine dynamics.

A good example is setting up a dial to function correctly. Take a dial powered by an AC motor through a clutch/brake. The clutch/brake is energized to spin a dial, which in turn advances the product on the dial from one station to the next. A proximity switch is used to sense when the dial has rotated enough to de-energize the clutch/brake to stop the dial. To set up the dial, the clutch/brake may need adjusting, but more importantly, the proximity switch needs adjusting. The dial can somehow be rotated into position, allowing the proximity switch to be adjusted. However, this doesn't take into account the inertia of the dial and is hard to perform. A better method is to provide manual-mode operations to start, stop, and jog the motor; engage and disengage the clutch and brake; and to single-cycle the dial. The dial can now be commanded to rotate and stop automatically after seeing the proximity switch. Additional logic is also added to interlock the manual-mode operations so the dial can only index when all the stations are clear of it and it is safe for the people working around it. The dial can now be operated just as it would when the machine is operating automatically at full speed, making it easy to adjust the proximity switch. The dial is commanded to index once, its location is checked after it stops, and then the proximity switch is adjusted as needed. After several trials, the process is done. The features added to adjust and check the dial are also useful for future maintenance because the clutch/brake will wear and require periodic adjustment.

To know what tools to incorporate into the machine logic, look at what is required to initialize each mechanism. In general, have the capability to command each motion. These same tools are also used to maintain the equipment, so incorporate the tools into the logic instead of using temporary logic. Temporary logic takes three times as long as logic that is planned out and added in advance. Temporary logic not only has to be removed, but it takes longer to write amid the confusion of writing logic at the machine instead of in a quiet office.

Maintenance also involves handling problems caused by worn out mechanisms, faulty sensors, out-of-spec component parts, and untrained operators. Detecting problems and providing methods to deal with them is called *fault recovery*. Fault-recovery logic can be some of the most difficult logic to write when the machine-control logic is not set up to accepts it. Not every problem can be anticipated, but general fault-recovery methods can be incorporated into the logic to deal with virtually every situation. Fault recovery starts by detecting trouble and is based on the fact that for every initiated action there are only two possible outcomes. One, the initiated action finishes as desired, or two, the initiated action fails. Feedback is important to detect failures. The more feedback provided, the quicker a problem can be detected and the more detailed the fault recovery can become. Any person who uses a personal computer is familiar with the blue screen of death. The computer locks up and requires re-booting to recover. The user has no knowledge of what went wrong nor is told by the computer of the problem. It would certainly be better if the computer could say why it is having problems. This would help pinpoint the source of the difficulty, so the problem can be fixed or avoided. When a pneumatic cylinder on a machine is commanded to extend, the proximity switch, indicating the cylinder is extended can be used to determine if the cylinder actually extended or not. If the cylinder failed to extend within its time limit, a message can be sent to the operator to say which cylinder failed, helping the operator to speed recovery.

Like the tools added to speed equipment startup and debug, the fault-recovery logic should be incorporated with the initial machine logic. Not only is it useful for maintenance to reduce downtime, but it is effective during startup. When initially operating the equipment by stepping through sequences, single cycling them, or dry cycling the machine, you might as well let the machine tell you what went wrong. After all, the fault-recovery logic needs checking, and you can start getting payback from the logic sooner.

Machine-Control Program Requirements

As already mentioned, a machine-control program has to sequence the equipment, but it also includes tools to assist in setting up and maintaining the equipment. What else does a program have to do and what other requirements does it have to satisfy? Knowing the requirements helps us build these features into the logic, and they are as follows:

Requirement 1 – Make Equipment Easy and Safe to Operate

The equipment must be safe to operate, and the machine-control program must work with the equipment's safety systems to help safeguard the people working with the equipment. As a result, I would place safety as the primary responsibility of the machine-control program. The logic needs to incorporate the following features to assist with safety:

- Stopping and removing power to prime movers
- Dissipating stored energy
- Complying with regulatory requirements
- Providing clear and informative messages

- Providing a consistent and intuitive operator interface

Requirement 2 – Sequence the Equipment

Of course, the machinery has to operate as desired to make good parts, and this is the primary intent of the machine-control program. To operate the equipment as desired, the machine-control programs needs the following:

- A clear definition of the machine's functionality
- Robust and reliable mechanisms

Requirement 3 – Minimize Commissioning Time

Starting up equipment for the first time is perhaps the most stressful time for a programmer. There is much to do, and there is a sense of uncertainty, especially with special equipment and processes that have never been built before. Types of problems encountered range from program typos to difficult processes. To help out, the machine-control program needs the following:

- Manual-mode layer
- Single-step, single-cycle, and/or dry-cycle modes
- Fault-recovery logic (error detection and recovery routines)
- Easy-to-edit logic (respond to changes faster)

However, no matter how much effort you put into a machine-control program, it can, at best, only complement a good mechanical design. The equipment may not work without the program, but it can work only as good as the design, selection, and fabrication of the equipment. The program itself cannot solve fundamental problems, such as undersized motors or flimsy frame members. Perhaps one of the best things a programmer can do is to understand the mechanical aspects of the equipment and know its limitations. Knowing when a problem cannot be solved through logic saves a lot of time.

Requirement 4 – Minimize Downtime

The end user can only realize the monetary investment in the equipment when it is producing. Thus the machine-control program has to minimize downtime to keep the equipment producing. Getting the equipment built, tested, and put into service faster also helps realize the investment. This is another reason to minimizing the commissioning time. To minimize downtime, the machine-control logic needs the following:

- Reliable logic
- Clear and informative messages for the operators
- Consistent and intuitive operator interface
- Fault-recovery logic
- Home all, purge, and other convenience functions
- Functions to help with calibration and other periodic requirements
- Manual-mode layer
- Single-step, single-cycle, and/or dry-cycle modes

Requirement 5 – Make Programs Easy to Change

Programs that are easy to change save time during the commissioning process and when the equipment evolves, and they help minimize downtime. Of course you do your best to anticipate and cover all the bases, but there is the inevitable mistake or difficulty with the process uncovered during equipment startup, which will require the logic to change. Logic that is easy to change allows you to adapt the logic faster so you can continue to move forward. After the equipment is placed into production, it may require changes to adapt to changes made to the processed product or even to a new product. There is also the gradual fine-tuning of the equipment to increase its quality and yield above the original equipment specification. This is especially true with special, one-of-a-kind machines that are built and tested in a limited amount of time. Special equipment typically needs the long-term running that production provides to fine-tune and optimize them to their fullest ability. Downtime is minimized when the maintenance people not only have the tools to efficiently work with the equipment, but also when they have a better understanding of the equipment. An easy-to-change program is also a program that is easy to learn and understand. We can summarize these machine program requirements as follows:

- Easy-to-read/understand programs
- Easy-to-edit programs

Requirement 6 – Make Programs Easy to Create

The last requirement is the ability to include all the above features and satisfy the above requirements in a timely manner. This can be accomplished with a standardized approach, where the basic structure of the logic is predefined and already includes the desired features. In theory, once you have defined what the machine is suppose to do, with all its requirements and constrained by its mechanics, the task of writing the logic is straightforward, simple, and even automatic. Actually, for many programmers the writing of machine-control logic takes on a building-block approach. After a programmer gains some experience writing logic for the typical machine mechanisms, the programmer develops a method of programming these devices and uses the same logic the next time he or she encounters these mechanisms. It is a natural process to use the lessons you have learned to save time the next time around.

Cascading Logic goes one step further and expands the building-block approach into a complete methodology. As you can see, many of the bullet points for the above requirements are common, thus showing that any methodology is finite and thus achievable. No methodology can completely include and foresee all the logical requirements of all types of equipment, but it can cover most of the requirements and it can provide a framework to get started. A programming methodology will save you time writing the bulk of the equipment's logic, thus providing more time for you to concentrate on the more difficult or unusual aspects of the equipment.

A standardized approach has other benefits besides saving time writing the initial logic. It also makes it easier to train others, like the end user and coworkers, on how the program works. A generic write-up explaining the logic can be made in advance and used as a training aid, or as I prefer, have them purchase this book. Coworkers will find it easier to lend a hand and be more proficient when they can acclimate themselves to your equipment faster. The training of new employees is faster and you can convey how you want programs written in a consistent manner. The end user can learn your machine faster and thus take better care of it and use it more productively. Therefore, the final requirement of a machine-control program is:

- A standardized approach

Chapter Summary

So what is the methodology of Cascading Logic? This will take the rest of this book to explain. The following summarizes each chapter.

Ladder Logic Basics. The main intent of this chapter is to make sure you understand how a programmable logic controller scans and solves its logic. It will do this by covering some basic ladder logic circuits.

Sequences and Motion Objects. This chapter illustrates a group of ladder logic rungs called a *motion object*. The motion object represents a single action of a machine. The motion objects are linked together to form a sequence where a sequence is an ordered list of steps the machine is required to perform, typically in a cyclic fashion. The rungs of a motion object initiate an action, determine when the action successfully finishes, and generate a fault if the action fails to finish. Motion objects provide an easy method of converting each step of a flowchart into machine logic.

Output Logic. This chapter explains how to link motion objects to physical outputs. Examples illustrate how a common block of rungs can control many diverse type of devices from solenoid valves to vision systems. The output logic also incorporates a complete manual-mode layer with bumpless transfer and interlocking.

Simulating Sensors. This chapter illustrates how to use timers in place of sensors to control machine activity. The examples further illustrate the usefulness of output logic.

Checking Sensors. This chapter examines the possible consequences when sensors fail and describes how to incorporate sensor checking into the machine logic. A common method of sensor checking, called *back checking*, is discussed, as well as *forward checking*, and a relatively new concept called *position bits*.

Controlling Sequences. This chapter focuses on how to start, hold, stop, and abort sequences to discuss the finer points of how sequences are controlled. The concepts presented here provide the basis for many advanced machine-control features.

Step and Single Cycle. This chapter expands the basic sequence-control concepts to provide two additional features. The first new feature is the ability to step through a sequence one step at a time. The second feature allows a sequence to be single cycled or operated once. Why you would incorporate these features and how they reduce downtime is also discussed.

System Control. Each machine section is controlled by its own sequence and these sequences are fairly independent. The system control ties all these independent sequences together to allow the machine to be started and stopped as a whole.

Parent/Child Sequences. This chapter illustrates how to control many sequences from one parent sequence. This method of sequence control is typically used with machines incorporating dials and walking beams.

Decision Logic. Many sequences perform different tasks depending on the outcome of a particular test. How a sequence decides which path to follow is determined by decision logic. This chapter will discuss decision logic and the general topic of program flow such as series and parallel sequence paths.

Dependent Sequences. This chapter will present methods to allow several sequences to talk to each other. The pitfalls of bad communications will also be discussed.

Machine States. This chapter illustrates how to characterize the operational status of a machine into predictable, defined states. The possible states include power off, fault, manual, not ready, ready, and running. The theory is that if you know the current state of the machine, you know how to navigate to the most desired state—the running state.

Message Logic. This chapter illustrates how to inform the operator of the machine status, warnings, and faults. The best way to reduce downtime is to provide the operator with timely and useful information.

1 Ladder Logic Basics

1.0 Introduction

This chapter will review ladder logic concepts by presenting and discussing several basic logic configurations and their application in machine control. The logic circuits presented here are simple and commonly used. Being able to recognizing these circuits and the functions they perform will help you in building logic and understand other people's logic. As a minimum, the reader should be familiar with programmable logic controllers in general, and the capabilities of their instruction sets. The ladder logic examples in Cascading Logic are based on the Allen Bradley SLC 500 instruction set. In my work I use many different types of programmable logic controllers, depending on my customers' requirements. I have found the specifics of the instruction set are not as important as the underlying concepts. Plus, the instruction sets are similar enough to allow conversion without much trouble. Once you start programming an application you will need to know the specifics of the chosen programmable logic controller. In particular you will need to know the following:

- How the inputs, outputs, and logic are scanned.
- How much time it takes to scan the I/O and logic, commonly referred to as scan time.
- How the memory is initialized. Are the bits and words reset to zero, random, or retentive?
- State of the outputs when the processor fails or when its operating mode changes.
- How the logic is constructed and executed.
- What instructions are available and how to use them.

1.1 Rungs and Branches

A rung is the smallest division of a ladder logic program; you can think of it as a single circuit, if you were constructing the circuit with discrete contacts, relays, and timers. A rung is made up of at least one input instruction and at least one output instruction. A typical input instruction is a normally open contact representing a physical input to the programmable logic controller or an internal memory element. A typical output instruction is a coil used to change the state of a physical output from the programmable logic controller or an internal memory element. The input instructions are arranged in combinations to create the desired logic for controlling the output instructions. Figure 1-1 shows a typical rung. Contacts in series such as contacts A and B form "AND" circuits. Contacts in parallel such as contacts D and G form "OR" circuits.

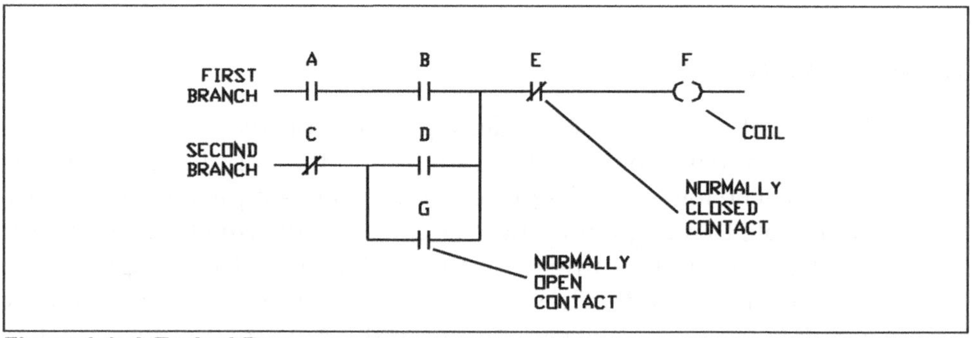

Figure 1-1. A Typical Rung

I use the word *branch* to refer to the distinct horizontal circuits of a rung. The rung in Figure 1-1 thus has two branches. In ladder logic, there are branch instructions used when creating the logical circuits. Figure 1-2 shows where these branch instructions are located.

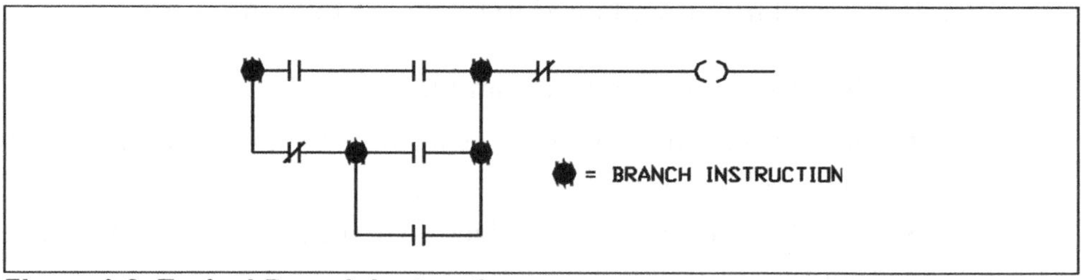

Figure 1-2. Typical Branch Instructions

1.2 Program Execution and Scan Time

Most programmable logic controllers execute their logic (scan) one rung at a time, starting with the first rung. After the last rung is executed, the scan starts over. The logic of each rung is resolved to update the output instructions appropriately. How the particular controller executes and resolves logic is important to understand because it affects how you construct your logic.

The controllers have other tasks to accomplish besides executing rungs. These tasks make up the total scan of the controller and include the following functions:

- Perform housekeeping functions and check for problems
- Read physical inputs and update memory accordingly
- Execute rungs
- Update physical outputs accordingly
- Perform network communication functions

The time the controller needs to complete its entire scan is called the *scan time*. The specifications for most controllers list their scan time in ms/k (milliseconds for each thousand words of executed ladder logic). The scan time listed is the time required to execute the ladder logic rungs, not the total scan time. The controller's literature should have a method of estimating total scan time. I find it takes more time for the controller to poll and update I/O than it does to execute the rungs.

The application of programmable logic controllers requires you to know the throughput time. On high-speed equipment, a sensor may change states so fast that the controller may not even see it toggling. Knowing the throughput time allows you to avoid this problem. When dealing with proportional, integral, and derivative (PID) loops, the controller's update rate can indicate whether you can control the process variable or not. I define the *throughput time* as the time between when a sensor is triggered to when a physical output of the controller changes state and includes the following:

- **Sensor Response Time.** The output signal from a proximity switch changes states after an object is placed or removed from the sensor's field. The time after an object is placed in the field to when the output signal turns on is the off-to-on response time. The time it takes the output signal to turn off after the object is removed is the on-to-off response time. A method to calculate the response time and signal duration of a proximity sensor triggered by a linear target and a rotating disk is shown in the Appendix.

- **Input Module Response Time.** Just like the lag in a sensor's output signal, an input module has time delays. The signal the controller sees lags behind the signal on the input terminals. This lag is worse for analog signals.
- **Controller Response Time.** This is the total scan time of the controller, not just the time to execute the rungs.
- **Output Module Response Time.** This is the delay in the signal the controller has sent to the output module and the signal on the output terminals. For instance, many alternating current output modules use silicone-controlled rectifiers (SCR). These devices change states when the voltage crosses zero volts. At best these devises can only switch every 8.33 milliseconds.

1.3 One Shots

The term *one shot* (o/s) is short for either a rising edge evaluation or a falling edge evaluation. In ladder logic, they are a pulse with a duration of one scan that occurs when a signal turns on (rising, leading, or on o/s) or when the signal turns off (falling, trailing, or off o/s). They are used to provide timing pulses or to give logic only one chance to evaluate.

A product inspection station uses a one shot to time when an inspection sensor is valid. A proximity switch monitoring a rotating disk provides the timing, and a photoelectric sensor looking at the product passing by performs the inspection. The output of the photoelectric sensor is constantly changing states depending on whatever it sees. Let's say the sensor is looking to see if labels are applied to bottles. The sensor's output will change states or see not only the edges of the label but also the gaps between the bottles. Instead of using a better photo-eye, we will use an on o/s pulse created from the proximity switch signal to tell us when to pay attention to the output of the photo-eye and when to ignore it. The disk has a feature that triggers the proximity sensor once per revolution (flag) and the disk is adjusted to turn the proximity switch on when the photo-eye is looking at the center of the label. The on o/s provides a specific signal that makes it easy to adjust the proximity switch. With the on o/s we do not necessarily care how long the proximity switch is on or when it turns off, only that the pulse occurs when the photo-eye is looking at the label.

Figure 1-3 and 1-4 shows how to create an on o/s and an off o/s. If the instruction set provides one shot instructions, then they should be used rather than programming two rungs of logic. However, the one shot logic illustrates two important concepts. The first is how ladder logic programs are executed, and the second is the importance of where rungs are placed relative to each other. The on and off o/s logic will not work if the execution order of the two rungs composing each circuit is swapped. Figure 1-5 analyzes the on o/s logic of Figure 1-3 rung by rung and scan by scan to illustrate how the logic evaluates into a pulse. I suggest you analyze the logic for the off o/s and the alternator circuits as a check of your program execution knowledge and to develop a technique for analyzing logic, since you will need one.

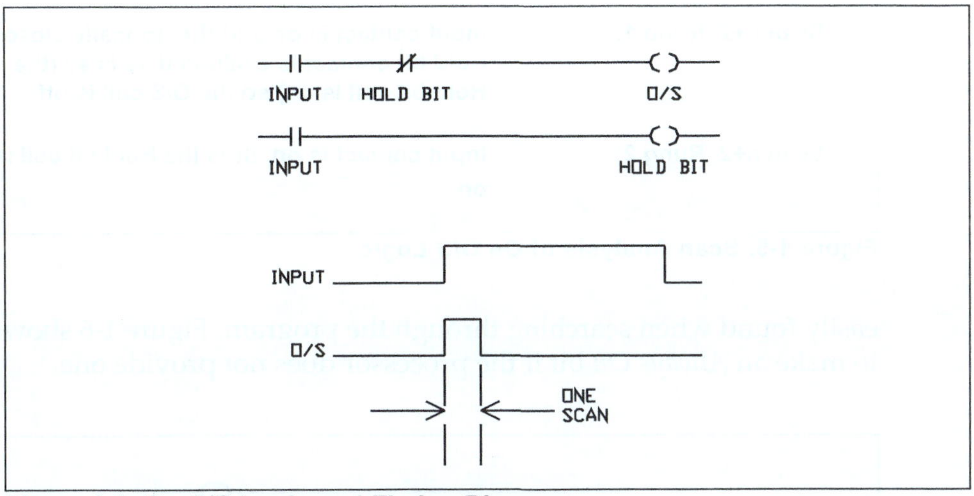

Figure 1-3. On O/S Logic and Timing Diagram

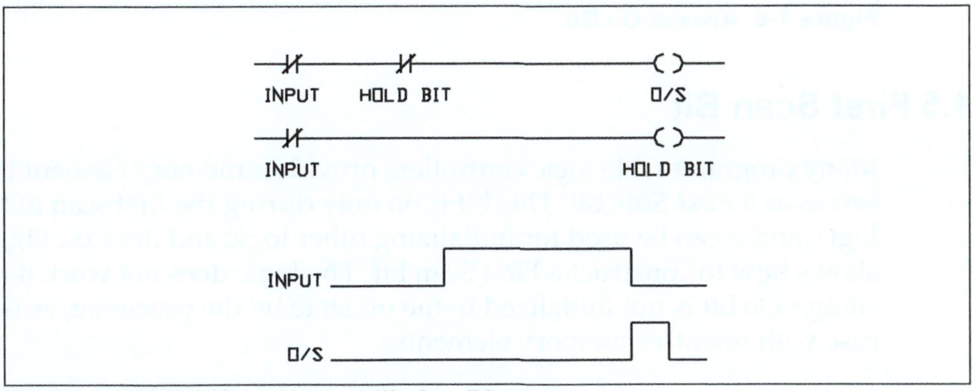

Figure 1-4. Off O/S Logic and Timing Diagram

1.4 Always On Bit

Having a memory element that is always on is useful when temporarily changing the logic. Since the *Always On* bit has its own description and unique memory element (address), it serves as a placeholder that can be

Scan n, Rung 1:	**Input contact is off, thus the O/S coil is off**
Scan n, Rung 2:	**Input contact is off, thus the Hold bit coil is off**
Scan n+1, Rung 1:	**Input is on and the normally closed Hold bit contact is evaluated as true (the Hold bit coil is off) so the O/S coil is on**
Scan n+1, Rung 2:	**Input contact is on, thus the Hold bit coil is on**
Scan n+2, Rung 1:	**Input contact is on and the normally closed Hold bit contact is evaluated as false (the Hold bit coil is on) so the O/S coil is off**
Scan n+2, Rung 2:	**Input contact is on, thus the Hold bit coil is on**

Figure 1-5. Scan Analysis of On O/S Logic

easily found when searching through the program. Figure 1-6 shown how to make an *Always On* bit if the processor does not provide one.

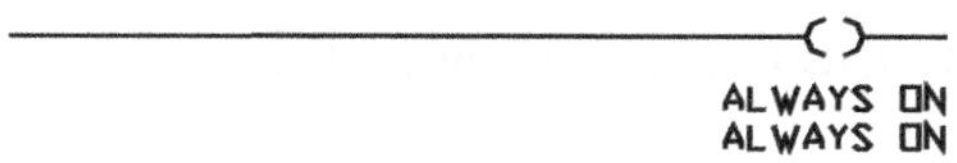

Figure 1-6. Always On Bit

1.5 First Scan Bit

Many programmable logic controllers provide a memory element that serves as a *First Scan* bit. This bit is on only during the first scan of the logic, and it can be used for initializing other logic and devices. Figure 1-7 shows how to construct a First Scan bit. The logic does not work if the *Always On* bit is not initialized to the off state by the processor, as is the case with retentive memory elements.

1.6 Alternators

Alternators are flip-flop circuits that change states on each occurrence of an input signal. A typical alternator circuit is shown in Figure 1-8. A good use of alternating circuits is to provide equal run time on backup equipment. In a fluid pumping application, a second pump is installed and ready to operate in case the primary pump fails. Instead of letting the

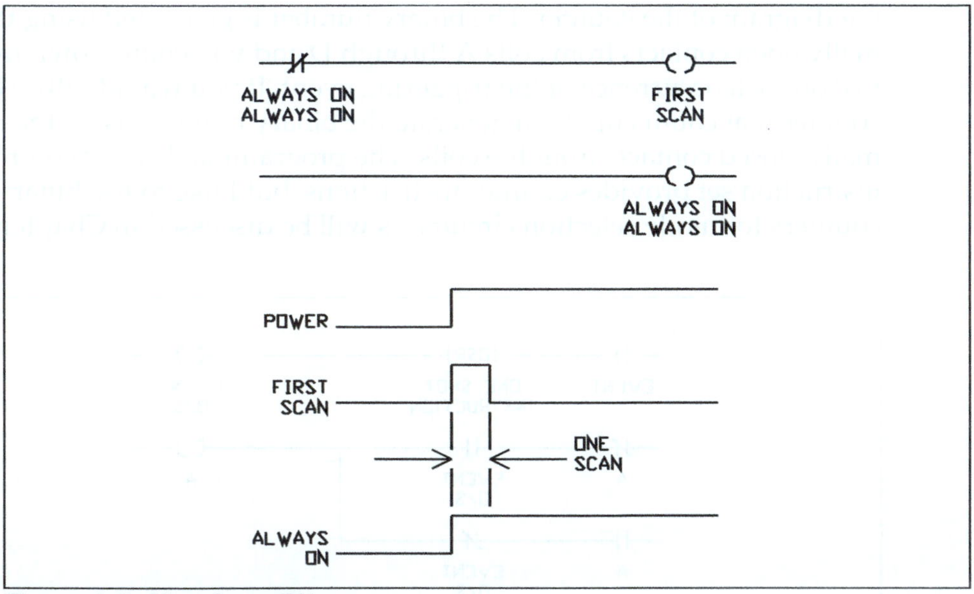

Figure 1-7. First Scan Bit

backup pump sit idle for years and rust shut, its usage is alternated with the primary pump. On each demand of the pumping system the last pump not used is energized, thus providing equal run time on each pump.

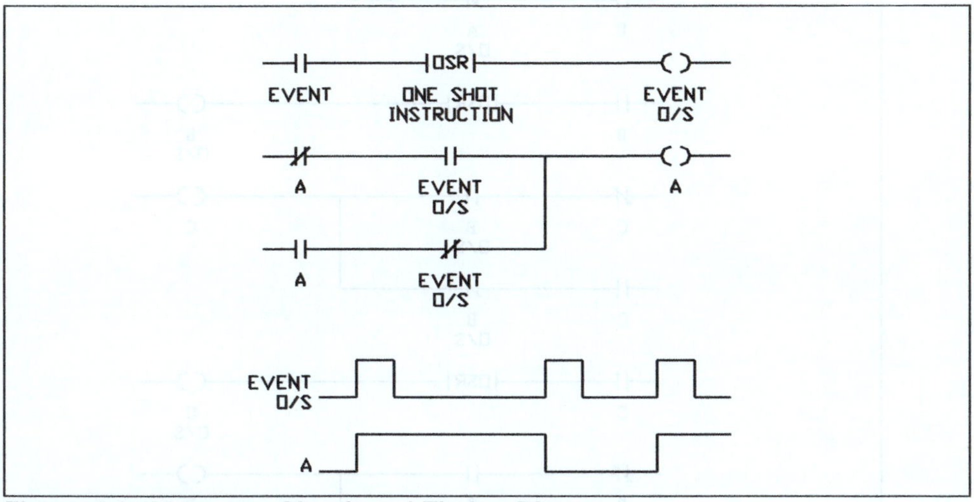

Figure 1-8. Alternator Circuit and Timing Diagram

1.7 Binary Counters

A binary counter can be made using successive alternator circuits, one alternator for each required bit. If you want to count between 0 and 15, use four alternating circuits as shown in Figure 1-9. Figure 1-10 shows the timing diagram of the counter. The binary number is generated using the normally open contacts from coils A through D and will count down from 15 to 0 on each occurrence of the triggering event. If you would rather have a counter that counts up, then generate the binary number using the normally closed contacts from the coils. The programmable logic controller's instruction set provides counter instructions, but I like to use binary counters for mode selection circuits, as will be discussed in Chapter 8.

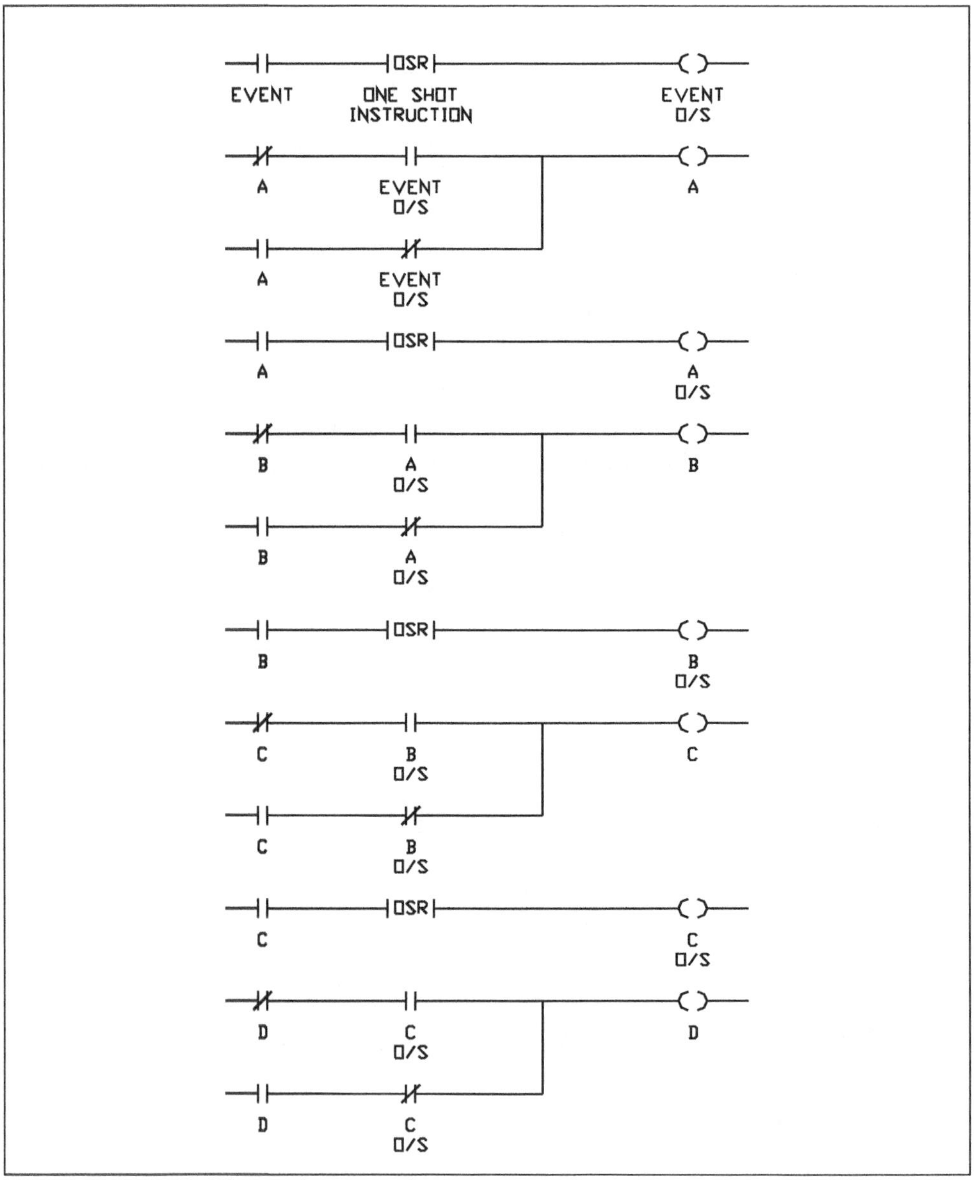

Figure 1-9. 4-Bit Binary Counter Circuit

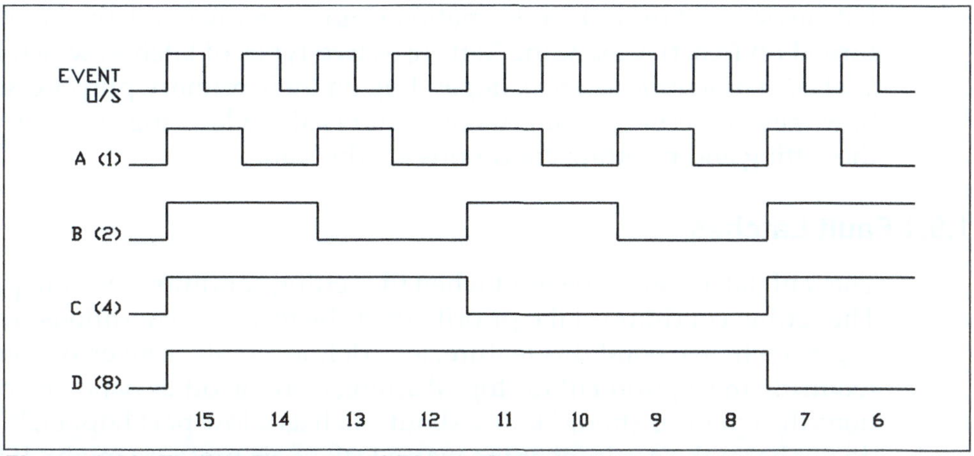

Figure 1-10. Timing Diagram of Binary Counter

1.8 One Scan Oscillator

The output of a one scan oscillator circuit toggles between on and off for each successive scan of the logic. The circuit and timing diagram is shown in Figure 1-11. One scan oscillator circuits are useful to provide pulses for other counter circuits and help to illustrate the execution and scan of the controller.

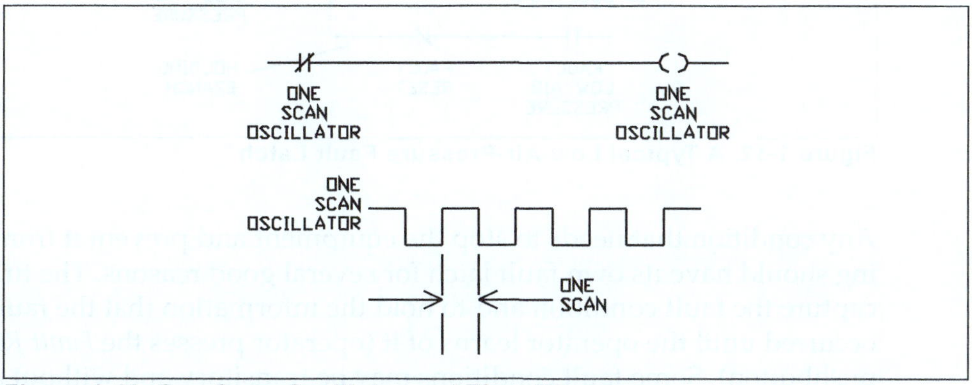

Figure 1-11. One Scan Oscillator and Time Diagram

1.9 Latches

A latch is a logic circuit used to hold one piece of data. Latches can be used to capture and hold an intermittent signal, to remember if a component part is good or bad, and to track and remember the equipment's progress through its cycle. The latch waits until one or more conditions are true and then turns on. You could say, when the setting conditions are valid the latch is set. Once the latch is set, it stays set, even if the setting conditions are no longer valid. The latch stays set with the help of a holding branch.

Latches stay set until the information is no longer needed, then they are turned off (reset). Cascading logic uses two types of latches, which I call a *fault latch* or a *sequence latch,* depending on their primary purpose and on how they are reset. The alternator circuit is also a latching circuit, where the setting and resetting conditions are the same.

1.9.1 Fault Latches

The fault latch cannot be reset when the setting conditions are still present. The setting conditions take priority over the resetting conditions. To explain why we want this feature, let's define a fault as an error condition requiring the equipment to stop. Warnings, on the other hand, are conditions the operator should know about, such as a low part hopper, but they do not harm the equipment the way a lack of air pressure might. If the problem is so bad the equipment should be stopped, then we also want to prevent the equipment from starting when there is a fault condition. As a result, we do not want to reset a fault latch unless the faulting condition is gone. A typical fault latch used to monitor a low-air-pressure condition is shown in Figure 1-12.

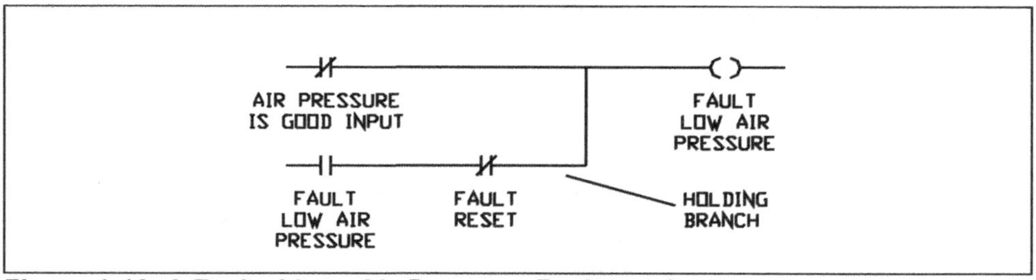

Figure 1-12. A Typical Low-Air-Pressure Fault Latch

Any condition that needs to stop the equipment and prevent it from starting should have its own fault latch for several good reasons. The first is to capture the fault condition and to hold the information that the fault occurred until the operator learns of it (operator presses the *Fault Reset* pushbutton). Some fault conditions may be transitory and without a latch to capture it, the machine would stop and you would not know why. By the time you look for the problem, the fault condition may have recovered. Air pressure is a good example of an intermittent problem. As the equipment cycles, the demand for air changes, causing the air pressure to fluctuate. After the machine stops, the air is no longer needed, so its pressure returns to normal. A second reason to have a separate fault latch is that they help to differentiate the problem. Instead of wondering what the problem might be, the equipment can simply say it is the air pressure, or it is a particular cylinder. This speeds recovery and reduces downtime.

1.9.2 Fail Safe

Notice in the fault latch of Figure 1-12 that the *Air Pressure is Good* input uses a normally closed contact. This is because the input is energized whenever the air pressure is good. The input de-energizes not only when the air pressure is low, but also when there is a problem with the switch's wiring (wire break or short circuit), or when there is a problem with the power supply feeding the circuit. As a result, the fault latch not only detects a low-air-pressure condition but also problems with the switch. This is fail-safe wiring; it cannot detect all the possible problems but it is better than the alternative. Look at the opposite case where a bad condition energizes the input rather than de-energizes it. If the wiring to the input breaks, then the processor will never see the bad condition should it occur.

1.9.3 Sequence Latches

The sequence latch can be reset at any time, even when the setting conditions are still present. The resetting condition takes priority over the setting conditions. A typical sequence latch is shown in Figure 1-13. Sequence latches build the sequences that tell the equipment's mechanisms precisely what to do. They remember where in the cycle the machine is, what is left to be done, and if the product in the machine is good or bad. When the equipment finishes a cycle, all the sequence latches are reset. The sequence may also be reset if there is a problem, and it is necessary to abort the sequence. The latches that make up a sequence need to be reset unconditionally, so the sequence latch is chosen for building sequences. Chapter 2 discusses sequences and sequence latches in more detail.

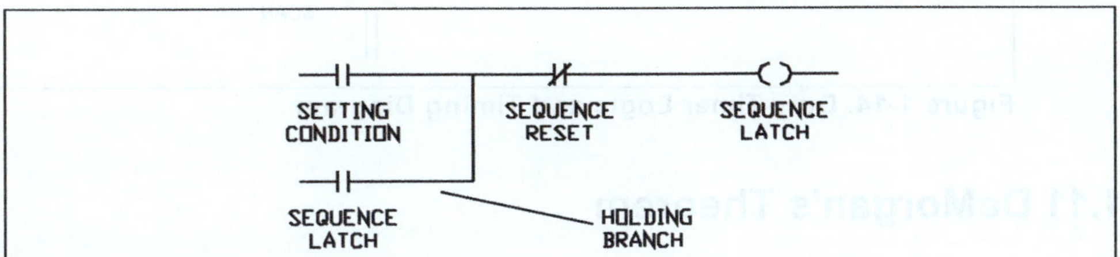

Figure 1-13. A Typical Sequence Latch

1.9.4 Latch Instructions

Programmable logic controller instruction sets provide set and reset instructions, commonly called S, R, Set, Reset, L, and U instructions. Retentive memory is another method of providing the functionality of a latch. Retentive memory has memory elements that maintain their on state, off state, or value when the programmable logic controller's power is cycled, or when the mode is changed. The latch instructions can certainly be used instead of the latch rungs. However, the latch rung is used in Cascading Logic to assist in illustrating the programming concepts.

1.10 Duty Timers

A duty timer circuit outputs a signal that constantly changes states. It is on for a certain amount of time, off for another time period, and then repeats. The duty timer circuit and timing diagram are shown in Figure 1-14. Duty timers are useful to blink overhead stack lights and pilot lights. You can make all kinds of timer circuits, but I find the duty timer the most useful because the on duration can be set differently than the off duration, even though I usually set both to one second.

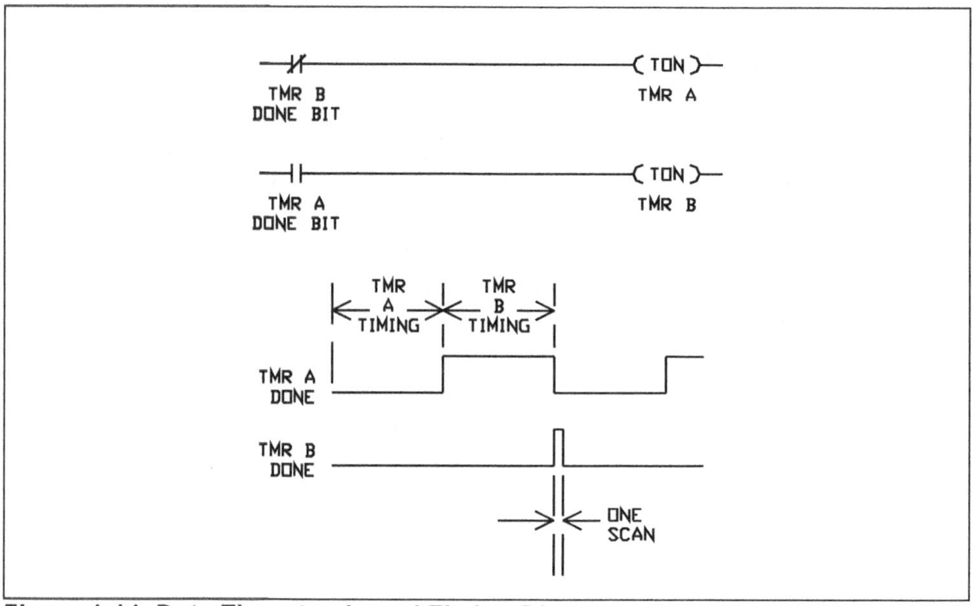

Figure 1-14. Duty Timer Logic and Timing Diagram

1.11 DeMorgan's Theorem

There are many instances when we desire to invert the output of a circuit. Instead of having the logic evaluate to a true state, we would rather have it evaluated to an off state or visa versa. Sometimes it is easier to think of a logic circuit one way, but find it easier to use the opposite elsewhere in the program. DeMorgan came up with an easy way of inverting logical circuits. In a nutshell, do the following to invert the output of a circuit as illustrated in Figures 1-15 to 1-17:

- Convert AND circuits to OR circuits
- Convert OR circuits to AND circuits

- Convert normally open contacts to normally closed contacts
- Convert normally closed contacts to normally open contacts

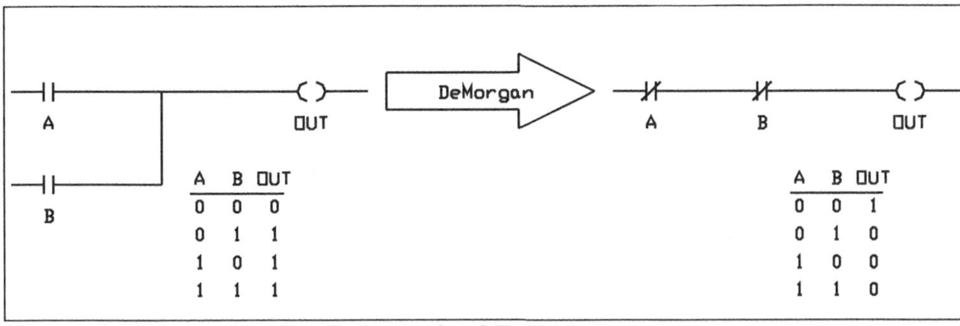

A	B	OUT
0	0	0
0	1	1
1	0	1
1	1	1

A	B	OUT
0	0	1
0	1	0
1	0	0
1	1	0

Figure 1-15. Inverting the Output of a OR Circuit

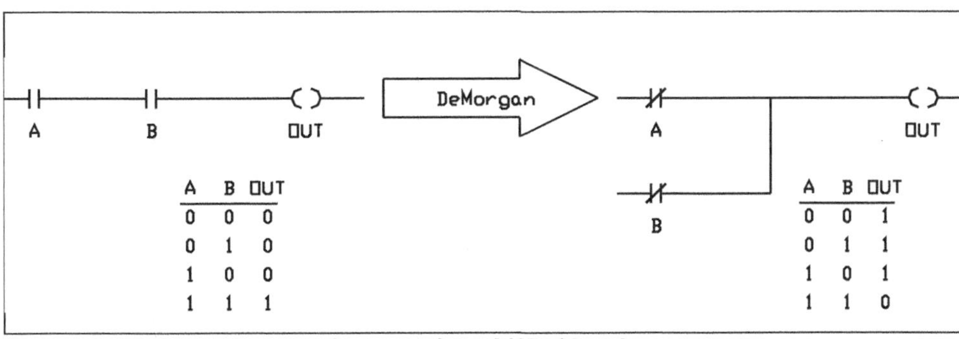

A	B	OUT
0	0	0
0	1	0
1	0	0
1	1	1

A	B	OUT
0	0	1
0	1	1
1	0	1
1	1	0

Figure 1-16. Inverting the Output of an AND Circuit

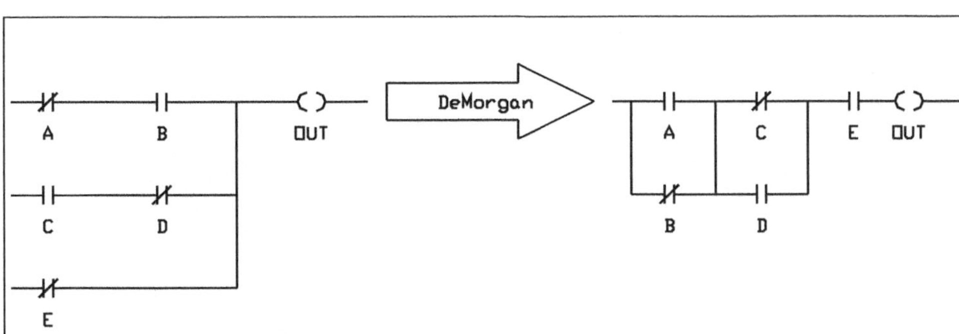

Figure 1-17. Inverting the Output of a Complicated Circuit

One thing to watch for is that the DeMorgan conversion does not work on logic circuits where a contact from the output coil is used in the circuit, such as latch and alternator circuits.

1.12 Other Circuits

The Appendix illustrates other useful and common logic circuits used for machine control.

2 Sequences and Motion Objects

2.0 Introduction

The focus of this chapter is to explain what a sequence is, how sequences are used to control mechanisms, and how to construct sequences. The chief ingredient of a sequence is what I call a motion object. *Motion objects* constitute each task or step the sequence performs and are linked together to build the sequences. The control of sequences, such as starting and stopping them, linking them to physical outputs, and linking them to other sequences is covered in the chapters to follow.

2.1 Functional Grouping

To make it easier to write machine-control programs, we first break the program into smaller, more manageable pieces. Each piece of the program is solved individually, allowing us to focus our attention on it. The programming objective is finished when the pieces are assembled back together into a complete machine-control program.

The disassembly of a machine for control purposes follows the functionality of the equipment. All the mechanisms on the equipment can be grouped on the basis of how they work together to perform a particular function. A small machine may have only a single group of mechanisms to perform a single function. The steps this function could have are shown in Figure 2-1.

Larger machines will have multiple groups of mechanism, where each group performs a single function. Figure 2-2 groups the mechanisms of an assembly machine into stations. Each station will have its own list of steps to perform, similar to the list of Figure 2-1. The operation of the assembly

1.	Wait for a work piece from the operator
2.	Wait for the operator to clear the work area
3.	Position and clamp the work piece under the tooling
4.	Punch a series of holes in the work piece
5.	Unclamp the work piece
6.	Eject the work piece

Figure 2-1. A Typical Function

machine is to execute the logic for station X, wait for the component parts to transfer, then execute the logic for stations 1 through 8 simultaneously. The programming effort is simplified by grouping the mechanism because now, instead of writing one big program, we can write nine smaller ones and then combine them into one. It is the divide-and-conquer approach.

Station 1.	Receive component A from a vibratory feeder bowl and place it into a fixture
Station 2.	Receive an O-ring from a feeder bowl and place it into component A
Station 3.	Receive component B from a feeder bowl and place it on top of component A
Station 4.	Ultrasonically weld components A and B together
Station 5.	Pressure test the welded assembly
Station 6.	Remove assemblies that failed the pressure test
Station 7.	Remove assemblies that passed the pressure test
Station 8.	Make sure the fixtures are empty for station 1
Station X.	Transfer mechanism moves fixtures and thus the component parts from one station to the next

Figure 2-2. Grouping Mechanisms into Stations

After the equipment is broken up into functional groups or stations, we can look at each one and break it into its individual pieces or steps. Figure 2-1 lists the individual steps of a function, and we can do the same for each station of the assembly machine listed in Figure 2-2. For each step of a function, we will write a motion object to control the particular step and then link these motion objects together to form a sequence. The sequences are then linked to physical outputs to complete the function. Latter we will discuss how to control the sequences and link them to other sequences to create a whole machine-control program. First, before we can start writing sequences, we need to organize ourselves and determine what each sequence needs to do.

2.2 Documentation

What we need to do before we can write a sequence for a group of mechanisms is to know everything there is about the mechanisms and the functions they perform. Figure 2-3 lists some typical items we need to know before we can continue.

- How each mechanism is controlled, such as the type of solenoid valves, motion controllers, relays, or serial interface
- The feedback from each mechanism, such as proximity switches or load cells
- What conditions are required to be satisfied before each mechanism is commanded to move, the order each mechanism is required to operate, and the decisions the functions will be making.
- The amount of time each mechanism has to operate
- Requirements for setup, adjustment, and calibration
- Identify risk area that may come back to haunt you

Figure 2-3. General Functional Documentation Requirements

Depending on the requirements for the particular application, the documentation can be anything from informal notes to formal write-ups and diagrams, but all the information needs to be known. Good documentation makes it easier for you to organize yourself and makes it easier for the end user to learn how to operate the equipment. A more informed end user has less downtime and uses the telephone less. Figure 2-4 list some typical forms of documentation used to express how the machine is to operate. The Appendix also has documentation for a pick/place mechanism to give you a flavor of some typical document forms.

2.3 Sequences

A sequence is an ordered list of steps that a group of mechanisms is required to perform. Sequences encompass the timing of the mechanisms and what conditions are required before each step is executed. Small machines that perform a single function will have only one sequence, whereas larger machines with multiple stations will have one or more sequences to control each station. The sequences follow the flowcharts or textual documentation and include all the minute detail necessary to perform the desired function. Sequences have definite requirements and these requirements are listed in Figure 2-5.

Electrical, Pneumatic, and Hydraulic Diagrams. These diagrams show how everything is connected and how the prime movers are coupled to the programmable logic controller. They should show amperages; analog scaling, such as 10 volts equals 5,000 pounds; servo motor scaling, such as 1 revolution equals 0.01 inches; cylinder bore and stroke sizes; and so forth.

Flow Charts. These charts are graphical representations of how the equipment is to function. They illustrate such things as what conditions are required to be satisfied before each mechanism is commanded to move, and the order each mechanism is required to operate.

P & ID Diagrams (Process and Instrumentation Diagrams). These diagrams show the prime movers, mechanisms, and feedback devices logically grouped per machine function to indicate cause and effect and how the control loop sees the equipment. Scaling and other useful information should be shown on these diagrams.

Theory of Operation and Control Function Description. These documents are textual descriptions of how the equipment should operate and express the same information as flow charts.

Timing Diagrams. These diagrams show when prime movers and mechanisms operate and how much time they have to accomplish their tasks. A must for all applications.

Operation Manuals. These manuals explain how to operate the equipment and document procedures for the safety systems, starting, stopping, filling hoppers, how the overhead stack lights work, and so forth.

Maintenance Manuals. These manuals explain how to set up, calibrate, adjust, troubleshoot, and maintain the equipment.

Figure 2-4. Typical Forms of Documentation

Specific Task. The sequences are constructed in a manner that allows the mechanisms to operate in a specific order and only in that order. They eliminate unforeseen combinations of logic to trigger events on the machine.

Cyclic. The equipment is cyclic if the mechanisms end up where they started from, so they can repeat their function on the next machine cycle. The sequences are constructed to allow the mechanisms to operate over and over again.

Features to Satisfy Other Programming Requirements. The introduction listed other requirements of a machine-control program. The sequences need handles and logic to accommodate these features. The features include error trapping, executing a sequence one step at a time, and so forth.

Readable. Because the equipment is sequential (does specific tasks in a specific order), the logic can be laid out to match, making it more readable and easier to edit. A well laid out and documented program goes a long way in assisting with debugging and training.

Figure 2-5. Sequence Requirements

A small code segment is shown in Figure 2-6 to illustrate the basic concepts of sequences. The segment shows the three steps that a pick/ place mechanism performs to grab and move a component part.

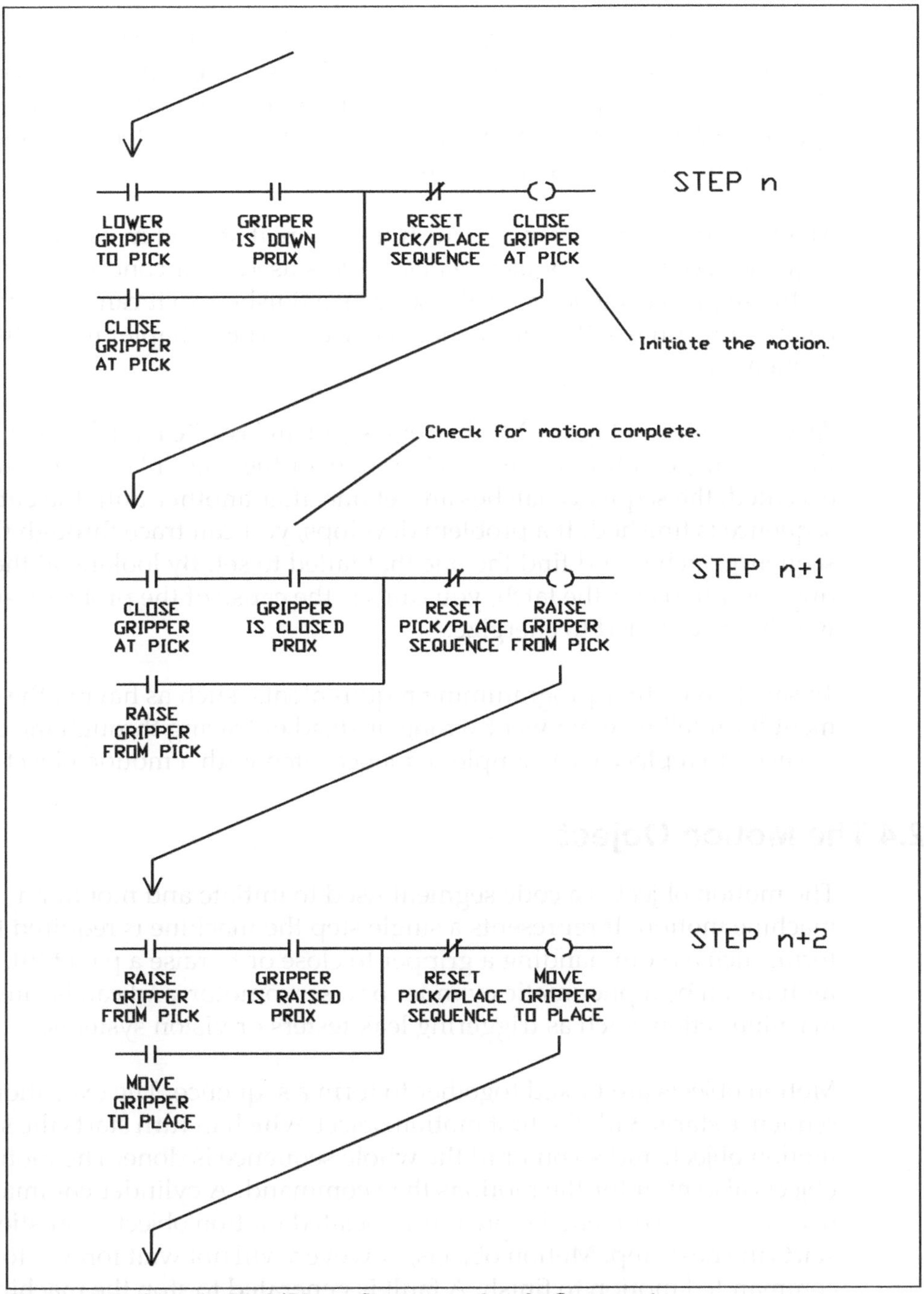

Figure 2-6. Simple Pick/Place Sequence Logic Segment

The three steps of Figure 2-6 use sequence latches to command each prime mover of the pick/place mechanism. Let's look at step n+1 to see how the logic meets most of the sequence requirements. The logic performs a specific task by making sure the last step is finished before continuing. The contacts *Close Grippers At Pick* and *Grippers Are Closed Prox* are the setting conditions for the *Raise Grippers From Pick* latch. Both of these conditions have to be satisfied before the grippers will rise. The contact *Close Grippers At Pick* is from the previous step and is used to keep each step locked in a specific order. The contact *Grippers Are Closed Prox* is the feedback that tells the logic the previous step has finished.

Making sure each motion returns from where it started makes the sequence cyclic. The contact *Station Reset* is used to unconditionally reset all the sequence latches after the sequence finishes, so it can start over again. More on resetting sequences is discussed in Chapter 6, Controlling Sequences.

As you can see in Figure 2-6, the code segment is quite readable. It also has the advantage of traceability. As the code for the pick/place mechanism is executed, the sequence latches are set one after another until the entire sequence is finished. If a problem develops, you can trace through the sequence latches and find the one that failed to set. By looking at the setting conditions for the latch, you can see the cause of the problem, such as which sensor failed to energize.

To satisfy the other programming requirements, such as having the equipment itself tell us what went wrong, instead of tracing through the code we need to replace each simple sequence latch with a motion object.

2.4 The Motion Object

The motion object is a code segment used to initiate and monitor a machine motion. It represents a single step the machine is required to perform, such as commanding a gripper to close or to raise a pallet lift. The motion can be a pneumatic cylinder or a servomotor, and can be any machine action, such as triggering leak testers or vision systems.

Motion objects are linked together to form a sequence. The execution of a sequence starts with the first motion object, which in turn starts the second motion object, and so on until the whole sequence is done. The motion objects also monitor the motions they command. A cylinder commanded to extend has to extend before the associated motion object is satisfied and starts the next step. Motion objects, however, will not wait forever for their commanded motion to finish. A fault is generated to stop the machine and alert the operator, if the commanded motion fails to do its job. Failures are typically detected by giving the motion a limited time to report success.

The motion objects and their grouping into sequences determine which motions move and when they move. It does not matter what form the physical outputs to the motions are in at this point, only the motion timing. The motion objects tell the output logic what it desires, and it is up to the output logic to do it. The separation of motion timing from motion output lets us concentrate on each part in turn. It is another divide-and-conquer method, allowing us to generate code faster.

2.4.1 A Typical Machine Step

A motion object is made up of three rungs of logic, called the motion initiation rung, the motion complete rung, and the motion failure rung. These three rungs for step n+1 of the pick/place mechanism, *Raise Gripper From Pick,* is shown in Figures 2-7, 2-8, and 2-9.

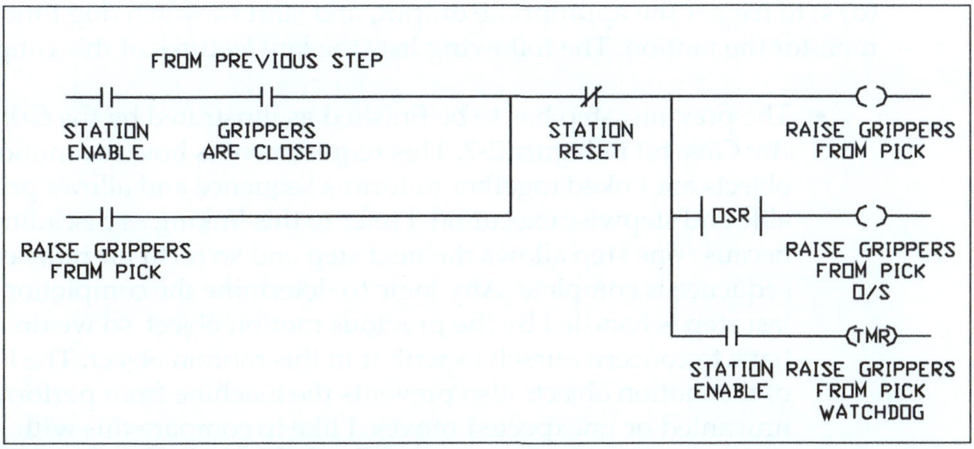

Figure 2-7. Motion Initiation Rung

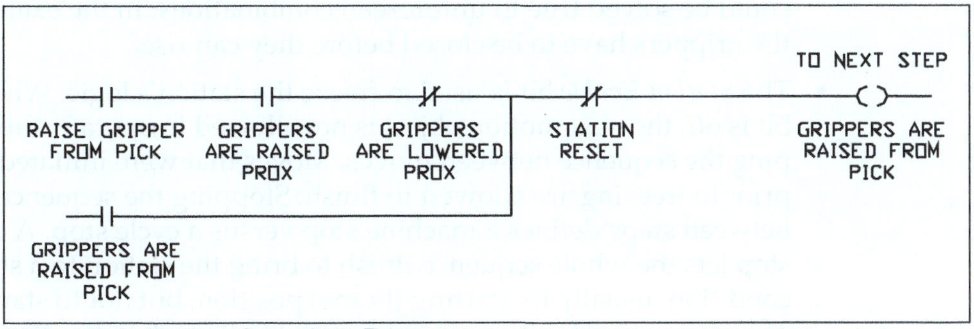

Figure 2-8. Motion Complete Rung

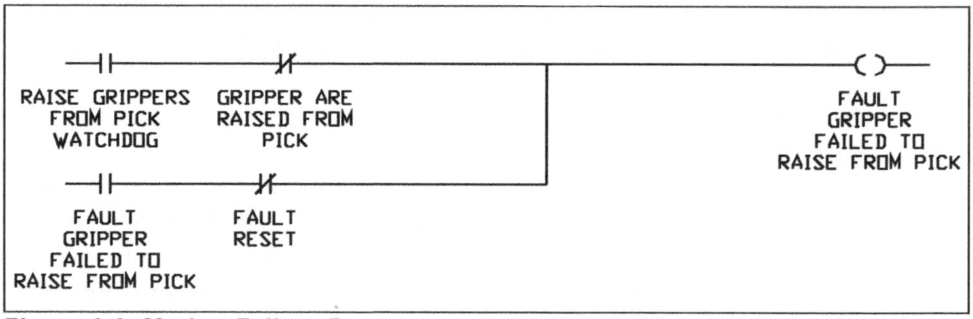

Figure 2-9. Motion Failure Rung

2.4.2 Motion Initiation Rung

The motion initiation rung is the first rung of the motion object. The motion initiation rung is shown in Figure 2-7. This rung, as the name implies, starts the motion. The motion is allowed to start provided the previous step finished and the station is enabled. Once both of these conditions are true, the motion initiation rung turns on, fires off a *One-Shot* bit (o/s) to trigger the appropriate output, and starts a watch dog timer to monitor the motion. The following lists the key features of this rung.

- The previous step has to be finished as illustrated by the *Grippers Are Close* bit in Figure 2-7. This requirement is how the motion objects are linked together to form a sequence and allows predictable and stepwise execution. I refer to this linking as cascading logic because one step allows the next step and so forth until the entire sequence is complete. Any logic to determine the completion of the last step is handled by the previous motion object, so we do not have to concern ourselves with it in this motion object. The linking of the motion objects also prevents the machine from performing unwanted or unexpected moves. I like to compare this with genetics because the order of execution is prescribed through the shear will of the programmed steps and not on collections of logic that could be solved true in unforeseen combinations. In the example, the grippers have to be closed before they can rise.
- The *Station Enable* bit is used to freeze the station's logic. When this bit is off, the next motion object is not allowed to operate, thus stopping the sequence between moves. Moves that were initiated just prior to freezing are allowed to finish. Stopping the sequence between steps defines a machine stop versus a cycle stop. A cycle stop lets the whole sequence finish to bring the station to a stable condition, usually its starting (home) position, but not to start over. The *Station Enable* bit can thus be used to allow stepping through the sequence for troubleshooting and setup. In step mode, all you

need to do is fire off the *Station Enable* bit for one scan to trigger the next motion object.

- The *Station Enable* bit is turned off if a fault is detected. The mechanisms and the sequence can now be viewed to see the conditions causing the fault. The ability to view the frozen logic is a big help while debugging the machine, especially if the logic has many decisions and branches to it. Freezing the logic also enables fault recovery. The fault can be reset, and the machine re-started after the station is inspected and the problem corrected. The station can now continue from where it stopped. The *Station Enable* bit is energized when the machine is started and stays energized until a fault is detected. Another latch, usually the first latch in the sequence and called *Run Station*, energizes when the sequence is started and stays on until the sequence finishes or is aborted. The *Run Station* bit remembers if the sequence was started, so the machine knows if the station is starting from its home position or going to resume from where it stopped.
- The motion initiation rung stays on until the whole sequence is complete. The *Station Reset* bit is fired after the last step is finished to clear all the latches in the sequence. The sequence can then start over and perform another machine cycle. Having the latches remain on as the sequence executes allows for tracing through the logic. The trail of set latches can be followed to the first unset latch to find problems and view the conditions causing the trouble.
- The *One-Shot* bit connects the motion object to the appropriate output, such as a solenoid valve or motor. What the *One-Shot* provides is a means of triggering a physical output or set of outputs to change states without maintaining a constant signal to it. This allows the output to receive triggers from many sources, without these sources interfering with each other. Typical sources include motion objects within the same sequence, from other sequences, and from the manual mode layer. In some cases, the status of the master control circuit is used as a trigger to handle power-up conditions. All the sources triggering an output need to coordinate their efforts. Linking of the motion objects in the proper order provides the needed coordination for sources stemming from the same sequence. A typical pick/place sequence will trigger the vertical motion of the grippers four times, twice to raise and lower the grippers at the pick position and twice to do the same at the place position. Handshaking between sequences is used to coordinate triggering of outputs for sources stemming from separate sequences, such as when moving parts between a station and a walking beam. Coordination of manual mode actions is accomplished by allowing only these actions when the machine is in man-

ual mode and not executing any sequences. In manual mode the sequences are inhibited from executing, thus preventing them from triggering outputs.

- The *One-Shot* bit separates the sequence logic from the output logic. One of the main concepts in separating the sequence logic from the output logic is to provide, to some degree, encapsulation. Separating the logic allows us to concentrate on the various parts, thus breaking the bigger task into smaller, more manageable objectives. At this point in logic development, we do not necessarily have to concern ourselves with physical outputs; our task now is to provide the triggers to fire the outputs. How we link the motion objects determines the order the outputs are fired. The appropriate logic for the outputs depends on the output type, the manual mode interface, and the desired action when motive power is either applied or removed. Solenoid valves, servo controllers, vision systems, and ultrasonic welders are some typical output types. Output logic is covered in the Chapter 3.
- The watch dog timer tracks the motion. Each motion has its own watch dog timer to provide timely reporting of error conditions. The alternative is one watch dog timer to monitor the entire sequence. The sequence watch dog is certainly easier to program, but its larger duration can be too excessive to monitor a single critical motion. The sequence watch dog is also harder to program if sequence freezing is added.
- Individual watch dogs can also be used to determine the time used to complete a particular move. This can be accomplished by either monitoring the watch dog timer while online or by adding code to capture the elapse time. The actual time needed to make a move can be compared with the timing diagram for machine debugging and cycle optimization. The logic needed to capture a motion's time is shown in the Appendix.

2.4.3 Motion Complete Rung

The motion complete rung, shown in Figure 2-8, is the positive acknowledgment that the motion completed its task. This rung also latches up until the entire sequence is complete to provide a solid reference point in the program. The benefits of the motion complete rung are listed below.

- It makes detecting problems easier by removing the conditions required to initiate the next step from the fault logic. The fault logic is simplified, making it easier to write and the program easier to read.

- It provides a clear stop point for stepping and fault recovery by allowing the insertion of a *Station Enable* bit in the next motion object.
- It keeps decision logic clean and free of confusion from the previous step. Decision logic is used to control how a sequence branches into multiple paths. When the next step in the sequence is a decision step, the decision can be made without additional confusion if a cylinder actually extended or not. This also makes the decision logic easier to read and write. Consider when a probe is extended to check for the placement of a snap ring. After the probe is extended, a snap ring sensor is interrogated. If the probe fails to extend, this condition can be detected separately from the inspection result of the snap ring being there or not. Decision logic is covered in Chapter 10.
- It stacks the deck in your favor. In many cases the same conditions used to signal completion are the same conditions used to start the next step, so why separate the rungs? Because it makes the program easier to write, debug, and change. The labor needed to write the program increases, but compare it with the labor spent in debugging when additional people, such as assembly personnel are standing around waiting for you. Labor costs increase dramatically again when the machinery is at the customer's facility holding up production. The additional time spent separating sequence steps is not all that much when you make it a standard concept, thus relieving you of the time needed to think about it. It also saves time in debugging by allowing clear divisions in the program to facilitate logic viewing, additions, and rearranging. Lastly, the motion complete rung makes the program easier to read and understand. This reason alone makes it worthwhile by reducing the number of phone calls from the customer.

The main reason for the motion complete rung is detecting the completion of the move. Figure 2-8 shows two proximity switches arranged in a back checking scheme to determine if the grippers lifted from the pick position. The first switch, *Grippers Are Raised,* is the sensor that actually tells the sequence the grippers did indeed rise. In a perfect world this would be the only sensor needed but sensors do fail so the second sensor, *Grippers Are Lowered,* is also included. The second sensor is checked to see if it is still functioning, it should be off at this point in the sequence.

Back checking is an attempt to detect a failed sensor before it can cause harm. Take the case where only one sensor is used, and this sensor failed on. The instance the motion is initiated it is also detected as done, and the next motion is immediately triggered. Instead of having the grippers of our pick/place example rise then move horizontally, it will instead do both simultaneously and thus damage itself. Back checking does not guarantee the prevention of short-circuited motions, but since the sensors are checked twice as often, the machine has a greater chance to detect a failure.

Forward checking is another method used to determine when a move is finished and to check sensor functionality. This method eliminates problems when sensors fail but requires more programming. Back checking and forward checking is covered in Chapter 5.

2.4.4 Motion Fault Rung

The fault rung shown in Figure 2-9 is the negative acknowledgment of the initiated move. This rung latches up if the motion failed to complete when the watch dog timer expires. The fault latch drops the *Station Enable* bit to freeze the sequence and most importantly to alert the operator. The resulting message to the operator can include what mechanism failed, where in the sequence it failed, the most likely sensor causing the problem, and a message number. The message for the fault can say "Fault 43 Station 3 Grippers Failed to Rise from Pick Check PRX 0347." The message number, 43, makes it easier to find additional information in the equipment manual.

The operator is needed to close the control loop when faults occur, but as a minimum, the fault latch can inform the operator of the situation. Operators can assess the situation better than the machine logic, and they have the ability to fix the problem. Involve the operators through messages and enable them to quickly re-start the machine by providing recovery options and a good equipment manual. Recovery options include trying again, aborting only the faulted station, and manual mode operations.

All faults have to be cleared before the machine can be started or restarted. The *Fault Reset* bit is used to break the holding branch of the fault rung, but it can only clear the fault if the fault conditions are gone. The conditions in Figure 2-9 are the motion complete bit, *Grippers Are Raised From Pick*, and the watch dog timer bit, *Raise Grippers From Pick Watch dog*. The motion complete rung may still be off, but the watch dog timer bit is off, so the fault can be cleared. The watch dog bit was forced off by including the *Station Enable* bit in the watch dog timer branch of the motion initiation rung in Figure 2-7. When the *Station Enable* bit turns off because of a fault, it will reset the watch dog timer.

One method of fault recovery is to re-start the machine and let the failed motion try again. Perhaps the motion failed because a sensor is out of position. The sensor can be adjusted and the machine re-started. When the machine is re-started and the *Station Enable* bit turns back on, the motion is given another time interval to complete. Giving the machine another chance allows it to handle simple problems without having to waste product or time. Fixable problems are called recoverable faults.

The fault latch provides the sequence with a definite outcome. Without the fault latch the program could hang between steps requiring somebody to plug into it with a support computer, a tool not associated with machine operators. The logic can also be reset by removing power to the logic controller, the so-called disconnect reset, but you will never hear the end of it if you use this method. Always having the sequences resolve themselves one way or another greatly increase the up time of the machine.

2.5 Summary

First we group the equipment's mechanism into function groups or stations and then determine the steps for each of these groups. A sequence is built for each group by programming a motion object for each step. When we program sequences we do not necessarily worry about outputs or how faults trigger operator messages. Those are different tasks to be done after the sequences are programmed. We concern ourselves only with the sequence steps. This allows us to do a better and faster job programming them. The motion object also frees us from the mechanics of programming through the use of a standardize approach. Once I have my flowchart, the programming almost falls completely into place.

3 Output Logic

3.0 Introduction

Chapter 2 presented sequences and motion objects. These control the order that the prime movers operate but do not include the logic for the outputs of the programmable logic controller. The motion objects contain only triggers to tell the outputs to change states and feedback to monitor their progress. This chapter will discuss the output logic needed to except the triggers from the motion objects and how to add a manual-mode interface. The output logic works with many types of prime movers, from pneumatic cylinders and servo controllers to vision systems and ultrasonic welders. The examples in this chapter focus on pneumatic cylinders for a pick/place mechanism coupled to the outputs by solenoid valves. Solenoid valves and cylinders constitute the majority of prime movers found on industrial equipment, and the outputs for these are the most common. See the Appendix for more information on a typical pick/place mechanism, and the plumbing involved.

The output logic uses three basic types of rungs called auxiliary rungs, output latches, and complementary outputs. Logic for virtually any type of output can be programmed with these three rungs. However, we need to consider the application before applying the appropriate output logic.

3.1 Application Considerations

In the application of logic to control an output, we need to consider what the output does, the effect of cycling motive power, and the effects of processor initialization. The following bullet points summarizes what has to be considered:

- **Master Circuit.** Safety circuits take many forms, but in general, they will be referred to as the *master circuit*. These circuits, external of the programmable logic controller, are used to control motive power for the safety of personnel. The prime movers are provided with whatever energy they need when the master circuit is energized. When the master circuit is de-energized, the motive power is removed to prevent the prime movers from working. Master circuits are de-energized when guard doors are opened, when emergency stop buttons are pressed, and when the processor is not executing its logic. The master circuit controls motive power with devices, such as contactors and air dump valves. A solenoid valve needs electricity, and the attached cylinder needs air pressure. Generally, both energy sources are removed from the cylinder when the master circuit is de-energized. With servo controllers, electrical power and the enable signal is removed from the drive, and brakes are engaged. Vision systems and similar equipment will remain on when the master circuit is de-energized because they do not release or store potentially harmful energy. Vision system strobe lights, however, can surprise an operator and can possible cause a head injury.
- **Motive Power.** Motive power is what makes the prime movers work and includes pneumatic pressure, hydraulic pressure, electricity and so forth. Another source of energy is gravity, which can move a mechanism after motive power is removed.
- **Removing Power.** We need to consider what happens to the prime movers and attached mechanisms when motive power is removed, and the key words here are kinetic and potential energy. The prime movers and attached mechanisms may still move after motive power is removed because gravity pulls on them or because of residual energy. Rotating and moving masses do not stop immediately because their kinetic energy and the pneumatic pressure take time to dissipate. A stamping press that is initially open when motive power is removed may fall slowly or suddenly because of gravity creating a pinch point beneath it. Stopping and preventing motion after motive power is removed involves adding brakes, cylinders with locking collars, and similar devices.
- **Restoring Power.** Another consideration is what happens when motive power is restored. The stamping press that fell because of gravity could now rise, creating a pinch point between it and an upper frame member. The state of the processor's output does not matter when motive power is removed, but it does when motive power is restored. The state of the outputs can be changed after motive power is removed to prevent motion, when motive power is

restored. The loss of motive power can be detected by monitoring the master circuit and pressure switches.

- **First Power.** Most programmable logic controllers reset their memory elements before executing their ladder logic. The exception is with retentive memory elements. Resetting of memory elements can occur after power is applied to the controller and when the controller is switched to run mode. You will need to understand how the processor you are using initializes its memory elements to apply output logic. The processor manuals should have information on how it initializes and when it occurs.

  The first power condition occurs when motive power is applied after the processor's memory elements have been reset. This is a special case of restoring motive power. The two main differences between restoring power and first power are the state of the outputs and the synchronization of the output states with the position of the prime movers and attached mechanisms. The state of the outputs with first power is typically off. If the processor has been running, then the outputs are in whatever states the logic has decided. The position of the prime movers and attached mechanisms will be assumed to be in an unknown position for first power. This is the worst case we want to plan for. If the power to the processor was off, then most likely it was off on the entire machine, and the processor does not knows the current position of the mechanisms. The mechanisms could have been moved by hand or could have fallen by gravity. If the processor has been running and motive power has been on, then the position of the devices controlled by the outputs and the state of the outputs will match. Gravity can, of course, make the devices differ from the state of the outputs, but this is predictable so we can plan for it.

- **Failures.** The type of failure we are concerned about with output logic is when the programmable logic controller fails. The motion objects in the sequences will catch when the prime movers and attached mechanisms fail. What concerns us is how the output modules respond when the controller fails. The state of the outputs could be de-energized, left alone, or even energized, although most outputs will de-energize. It is a matter of knowing how the outputs respond, so the effect on critical motions can be analyzed. Consider an output cycling on and off to proportionally control a heater. If the processor should fail, will the heater shut off or over heat?

3.2 Spring-Return Valves

The spool of a spring-return valve will shift when the solenoid is either energized or de-energized. The spool is the part of the valve that connects the ports of the valve together to control the direction of airflow through the valve. Figure 3-1 shows a schematic representation of a spring-return valve and a connected cylinder. The cylinder will return to its starting position when the output to the valve is de-energized. This is a good fail-safe feature to return a cylinder when motive power is removed, as it would when an emergency stop button is pressed. The pneumatic supply is also dumped when the master circuit is de-energized, but there is usually enough residual pressure to effect the move. A good example of using this feature is when a cutting tool is extended to cut into a rotating part. When the master circuit is de-energized, the tool will retract away from the part as the part starts to decelerate, thus avoiding harm to the tool.

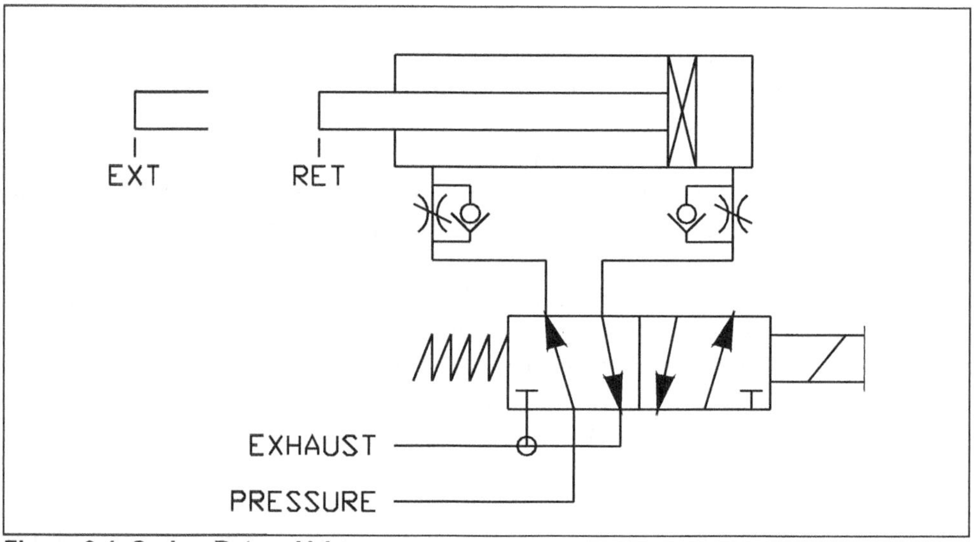

Figure 3-1. Spring-Return Valve

The next three figures illustrate the output logic needed to support the spring-return valve.

3.3 Auxiliary Rungs

Auxiliary rungs (aux) collect all of the triggers that are requesting the output or outputs to change states. The auxiliary rung in Figure 3-2 collects all the commands to lower the pick/place gripper. The auxiliary rung in Figure 3-3 collects all the commands to raise the gripper. Auxiliary rungs make the output logic easier to read and change. The triggers can come from many different sources, such as motion objects in sequences, homing routines, and the manual-mode layer. As Figures 3-2 and 3-3 show, the vertical motion of the pick/place mechanism receives four triggers from a sequence and two from the manual-mode layer.

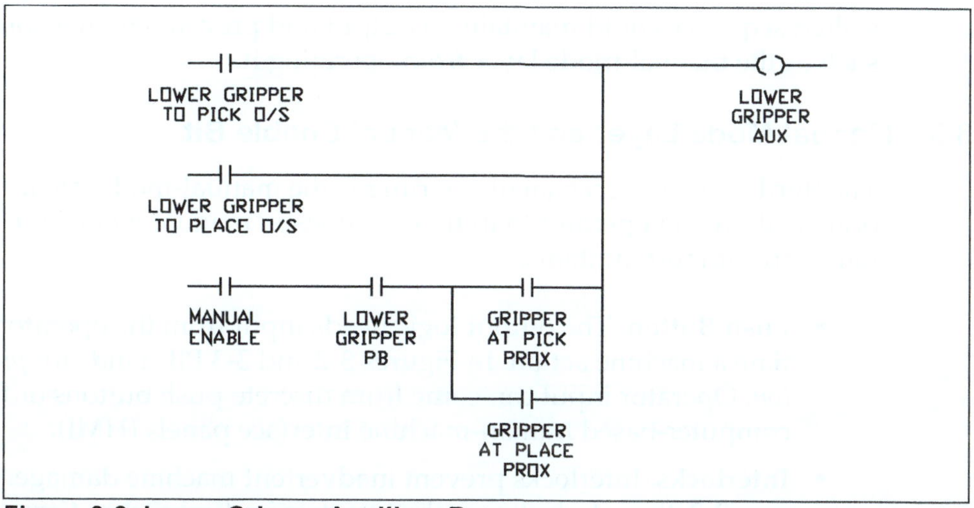

Figure 3-2. Lower Gripper Auxiliary Rung

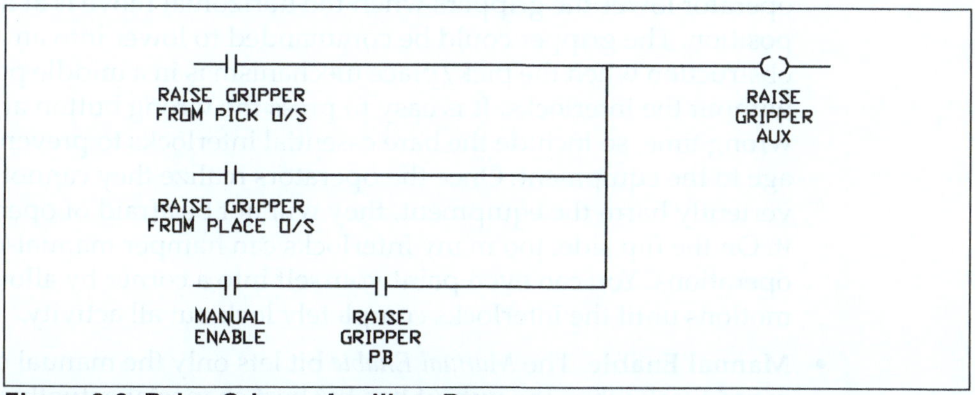

Figure 3-3. Raise Gripper Auxiliary Rung

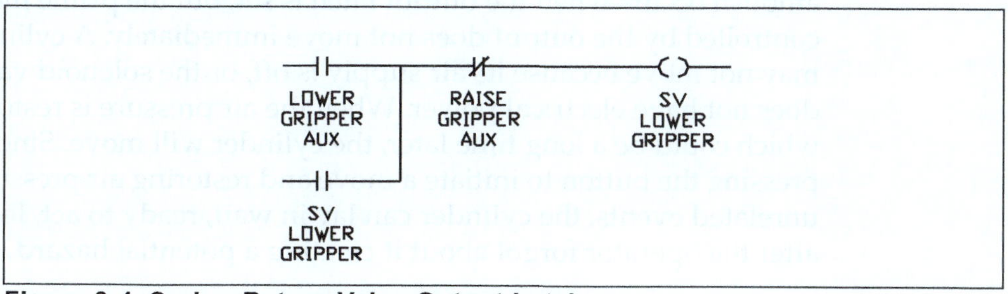

Figure 3-4. Spring-Return Valve Output Latch

3.3.1 Sequence Triggers

The first two branches of each auxiliary rung are triggers from motion objects. These are *One-Shot* (o/s) bits to let a sequence trigger an output but not to keep it in a particular state. Without the o/s bits, a frozen or stalled sequence could maintain an output and prevent another source, such as the manual-mode layer from operating it.

3.3.2 Manual-Mode Layer and the Manual Enable Bit

The third branch of each auxiliary rung is the manual-mode layer. This branch allows an operator to initiate motions in manual mode and contains three important items:

- **Push Button.** The output logic needs input from the operator to initiate a machine action. In Figures 3-2 and 3-3 PB stands for *push button*. Operator input can come from discrete push buttons or from computer-based human-machine interface panels (HMI).
- **Interlocks.** Interlocks prevent inadvertent machine damage. In Figures 3-2, the interlocks are the two proximity switches *Grippers at Pick Prox* and *Grippers at Place Prox*. These interlocks will only let the operator lower the grippers when the horizontal move is at an end position. The gripper could be commanded to lower into an obstruction when the pick/place mechanism is in a middle position without the interlocks. It is easy to press the wrong button at the wrong time, so include the bare essential interlocks to prevent damage to the equipment. Once the operators realize they cannot inadvertently harm the equipment, they will not be afraid of operating it. On the flip side, too many interlocks can hamper manual-mode operations. You can even paint yourself into a corner by allowing motions until the interlocks completely lock out all activity.
- **Manual Enable.** The *Manual Enable* bit lets only the manual command work when the output has the best chance of actually working. This is an important bit because it prevents ambushes. An ambush occurs when the output latch is set, but the prime mover controlled by the output does not move immediately. A cylinder may not move because its air supply is off, or the solenoid valve does not have electrical power. When the air pressure is restored, which could be a long time later, the cylinder will move. Since pressing the button to initiate a move and restoring air pressure are unrelated events, the cylinder can lay in wait, ready to act, long after the operator forgot about it creating a potential hazard.

Figure 3-5 shows the construction of the *Manual Enable* bit. This bit has to look at all of the conditions that could prevent the prime mover attached

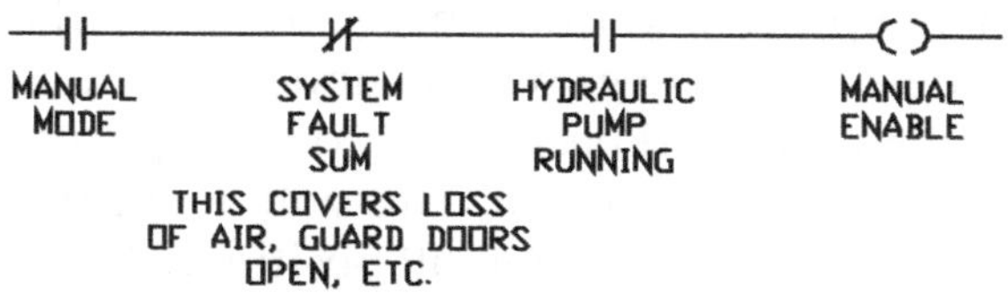

Figure 3-5. Manual Enable Logic

to the output from working. The *System Fault Sum* bit is a collection or summation of all the machine faults, and it is on when any fault is active. The system fault summation includes faults to indicate motive power is gone, such as the master circuit is off, or air pressure is lost. Figure 3-5 shows a bit for use with hydraulic cylinders that indicates that a hydraulic pump is running. A servo ready bit can also be inserted when the outputs are part of an interface to a servo controller.

3.4 Spring-Return Valve Output Latch

Figure 3-4 shows the physical output for the spring-return valve. The valve's output rung is a latch, which changes states per the requests of the auxiliary rungs. The three important things to remember about output latches are:

- **Bump-less Transfer.** The output latch will not change states when the machine's mode is changed from automatic to manual or visa versa, commonly called bump-less transfer. The output latches can change states only when told to do so, with the exception of being reset when the program is initialized.
- **Memory.** The output latch will remember the state of the output independent of the source or sources triggering it and independent of the motive power associated with the output circuit.
- **Default Position.** The default position of the prime move and attached mechanism associated with a particular output is defined as the position they are in when the output latch is off. The default position of the pick/place mechanism's vertical motion is raised. Knowing the default position helps when analyzing what happens during first power. The default position is generally picked as the best fail-safe position possible out of the choices available.

3.4.1 First Power

The output latch is reset when the processor is initialized. After initialization, the output latch only changes states when told to do so by the auxiliary rungs. The spring-return valve connected to the output starts up de-energized because the master circuit is also off. The master circuit is de-energized and prevented from energizing when the processor is not executing its logic. Ignoring gravity, cylinder motion will not occur when motive power is applied because the default position of the cylinder matches the state of the output. This will not be true when gravity pulls the cylinder away from the default position. The vertical motion in the pick/place example will rise on first power.

Motion on first power is not necessarily bad because first power cannot occur unless the guard doors and other safety devices are in place. The safety system has to be good before the master circuit can be energized. In the pick/place example, it might even be beneficial to have the gripper raise on first power. At this point it is not known if the horizontal motion is going to move or not. The gripper will most likely hit something if the gripper moves horizontally when it is in the down position. To help protect the gripper, get it up and out of the way as soon as possible by allowing it to move on first power.

Sometimes we want to prevent motion on first power when we know gravity or another influence will position a cylinder away from the default position. This can be accomplished by inserting a *First Scan* bit into the appropriate auxiliary rung, as shown in Figure 3-6. The *First Scan* bit presets the output latch in the first scan of the logic and before the master circuit has a chance of energizing.

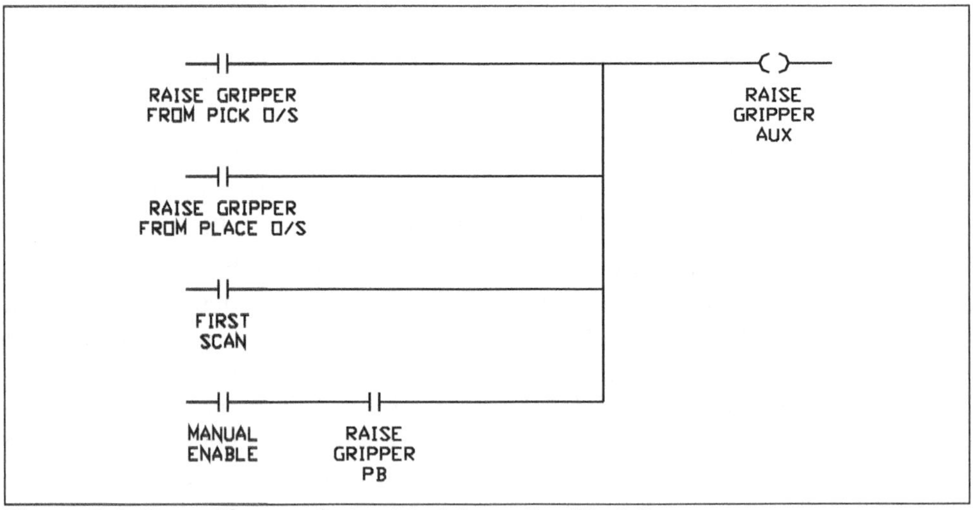

Figure 3-6. Presetting an Output Latch on the First Scan

3.4.2 Cycling Motive Power

The memory capability of the output latch is useful when motive power is cycled. The cylinder connected to the spring-return valve will be extended when the output latch is set. Loss of electrical power will de-energize the solenoid valve, causing the cylinder to retract. Loss of pneumatic power and gravity may also cause the cylinder to retract. The cylinder will be positioned back to its extended position when motive power is restored because the output latch remembered the desired position of the cylinder.

Sometimes we do not want motion to occur when motive power is restored. In the rotary cutting example, we do not want the cutting tool to extend into the part after restoring motive power—the tool could break. The output latch can be reset when motive power is lost, preventing the tool from extending after power is restored. This can be done by inserting the *Fault Master Is Off* bit into the output rung, as shown in Figure 3-7. When the master is de-energized, the spring-return valve will de-energize, causing the connected cylinder to retract and the output latch to reset. When the master is re-energized, the connected cylinder will remain in the retracted position.

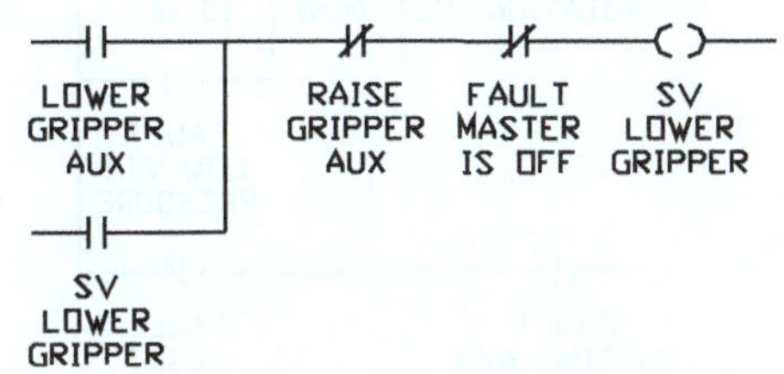

Figure 3-7. Using the Master Circuit to De-energize a Spring-Return Valve Output

3.5 Recovery

The next problem has to do with recovery. How can the controlling sequence resume operations after motive power is restored if a cylinder does not return to its rightful position? There are several ways of doing this.

One method is to reset the last motion object in the controlling sequence to effectively back up the sequence one or more steps. I find this an effective method when absolute servo moves are interrupted. Re-issuing the command tells the servo controller to go to position X, regardless of whether the servomotor has stopped short of, coasted past, or actually stopped at position X.

The whole sequence can also be reset to completely restart the task. In the rotary cutting example, it may be best to restart the entire sequence so the part can be brought back up to rotational speed and the tool's transverse motion retracted to the beginning of the cut, instead of leaving it in the middle where it was interrupted.

Programming for sequence recovery starts by creating a fault to detect when motive power is lost during critical sequence steps. A bit from the motion initiation latch of the first critical step is used to enable the fault. A bit from the motion complete latch of the last critical step is used to disable the fault. A fault that might be used for the rotary cutting example is shown in Figure 3-8. This fault can set only when motive power is interrupted when the part is being cut, not before or after the cutting portion of the sequence. All of the faults associated with a sequence will freeze the sequence to prevent it from initiating any new steps. This was mentioned in Chapter 2 and will be discussed in detail in Chapter 6 on controlling sequences. Active faults also prevent the equipment from restarting.

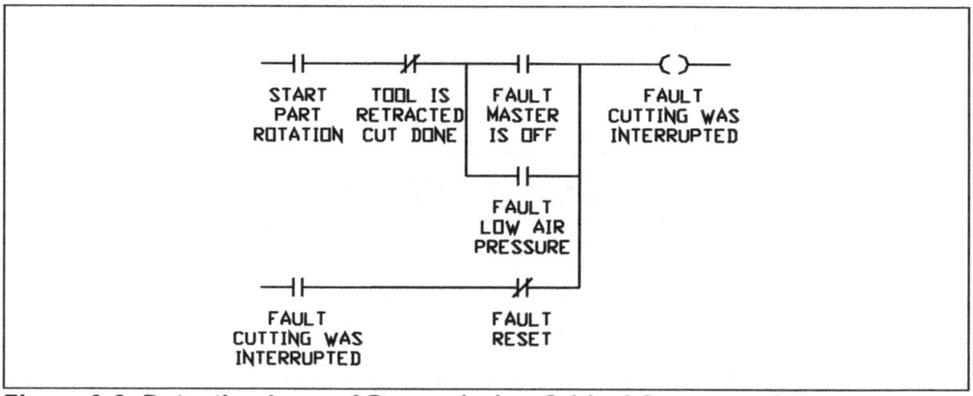

Figure 3-8. Detecting Loss of Power during Critical Sequence Steps

The sequence can be backed up by resetting the appropriate latches in the sequence, as shown in Figure 3-9. The whole sequence can be reset (aborted) by using the fault latch to trigger the *Station Reset* bit. The *Station Reset* bit is discussed in more detail in Chapter 6 on controlling sequences.

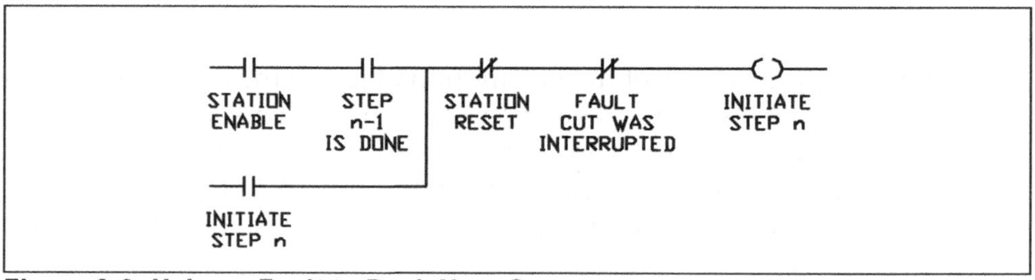

Figure 3-9. Using a Fault to Back Up a Sequence

Other output latches may also need adjusting to make the sequence ready to resume. It would be easy to correct these outputs as was done in Figure 3-7, but sometimes we cannot. In the rotary cutting example, the motion that transverses the tooling across the face of the part will need to be retracted. This cannot be done unless the tool is retracted away from the part, or we risk dragging the tool across the part and breaking it. We cannot do it when the fault is tripped because there is no motive power, and we do not want to create an ambush condition. The best time to do this is when the machine is restarted. Simply include a step in the cutting sequence to retract the transverse motion before the part is cut. This can be done while the part is accelerating up to speed. During normal sequence execution the transverse motion will most likely be retracted. Telling it to retract when it is already retracted will not hurt it. The motion object that tells it to retract will immediately see it is retracted and set the motion complete latch. The sequence then continues without missing a beat. During recovery, the sequence will retract the tool's transverse motion, accelerate the part, and then re-cut it.

Incorporating recovery features during the program-creation stage of a project is best, but it can be difficult with all the other aspects of the equipment to consider. Obvious problems can be planned for, but most other problems will appear when the equipment is commissioned. Adding recovery features at that time is not too difficult because the logic for the sequences, motion objects, auxiliary rungs, and outputs latches are flexible and easy to edit. When considering adding recovery logic, ask yourself the following questions:

- Is the component part worth saving? If not, then abort the sequence and get on with your life.
- Can downtime be reduced? The component part may not be worth saving, but if inconsistencies in the component part cause repeated trouble, then find someway of recovering quickly.

3.6 Double-Ended, Detented Valves

Figure 3-10 shows a cylinder connected to a double-ended, detented solenoid valve. The valve is called double ended because it has two solenoids. Energize one solenoid to extend the cylinder and energize the other to retract it. The valve is referred to as detented because its spool will remain in its last position if both solenoids are de-energized. Think of the spool as a memory element. As a result, the connected cylinder will also remain in its last position. Never energize both solenoids at the same time—the spool's movement, and thus the cylinder's, is unpredictable; plus the solenoids will melt. A loud buzzing sound should tip you off when this happens.

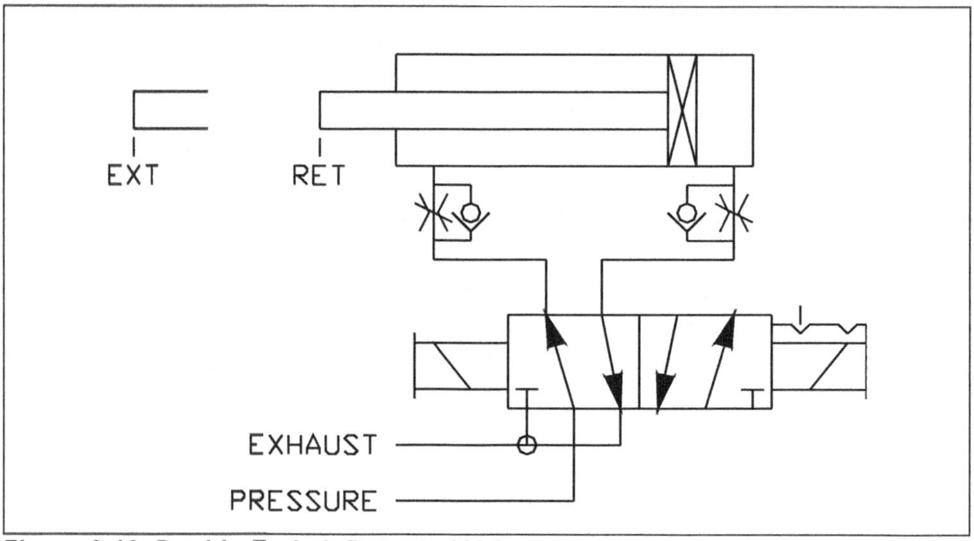

Figure 3-10. Double-Ended, Detented Valve

3.7 Double-Ended Valve Output Logic (Spring-Return Configuration)

The output logic in Figure 3-11 is for a double-ended solenoid valve treated as a spring-return valve. This configuration operates the valve just like a spring-return valve, but it allows us to pick the default position of the cylinder. In this example, the default position of the gripper is raised. When the processor is initialized, the *SV Lower Gripper* latch is reset, and the output *SV Raise Gripper* will be on. Notice that the logic prevents both solenoids of the valve from energizing at once, what I call push-pull outputs.

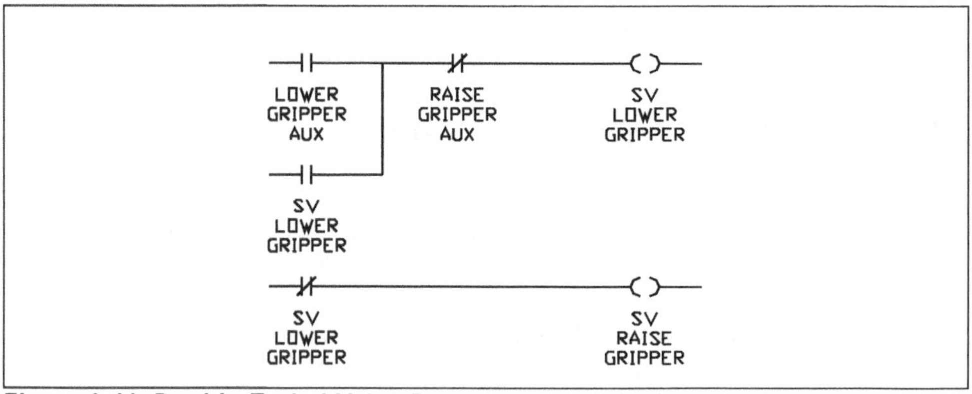

Figure 3-11. Double-Ended Valve Programmed as Spring-Return Valve

3.8 Double-Ended Valve Output Logic (Double-Ended Configuration)

To make full use of a double-ended valve, we need to program it as one. Each solenoid of the valve has its own output latch, as shown in Figure 3-12. Notice how the two auxiliary bits are used in each output latch to prevent both solenoids from energizing at once.

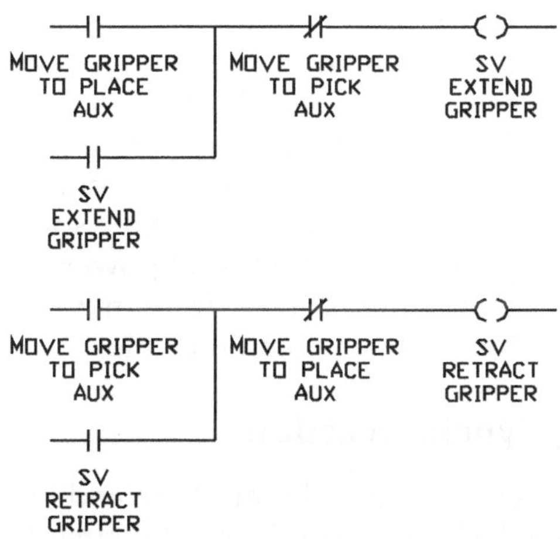

Figure 3-12. Double-Ended Valve Programmed as a Double-Ended Valve

3.8.1 First Power

The spring-return configuration creates a default position that the cylinder will go to on first power. The double-ended configuration creates no default position because both output latches are reset when the processor initializes. The position the cylinder goes to on first power is determined by the position of the valve's spool. The most likely scenario is no motion will occur on first power because the valve's spool and cylinder position are in agreement. They should be in sync, since the valve last commanded the cylinder and nothing has changed. Gravity, of course, puts holes in this theory and so does somebody moving cylinders while they are inside the machine or by operating valves by hand. If motion occurs on first power, other than because of gravity, at least the logic did not tell it to move; both output latches are off.

It is best to leave the horizontal motion on the pick/place mechanism in its current position on first power because we do not know the vertical position of the gripper. The gripper could be down because of gravity and movement on the horizontal axis could ram it into an obstruction. It is best to raise the gripper before moving it horizontally. If the current position of the horizontal move is wrong, the operator can use the manual-mode layer to reposition it or to home the entire mechanism. The machine helps by providing "not home" messages to alert the operator to misplaced mechanisms. The manual-mode layer can include a *Home Station* push button and/or a *Home All* push button, for convenience and to reduce downtime. The use of the auxiliary rungs makes it easy to include triggers from homing sequences.

3.8.2 Cycling Motive Power

The main reason for using a double-ended valve is to prevent the connected cylinder from moving when motive power is cycled. When motive power is removed, both solenoids of the double-ended valve will de-energize and leave the spool of the valve where it was last. The output latches also remember the desired cylinder position. When motive power is restored, no motion will occur because the solenoids will energize the same way they were before the loss of power, keeping the spool from moving and thus keeping the cylinder from moving. Of course, gravity makes a mess of this, but at least it is predictable.

3.8.3 Maintaining Synchronization

What if the operator enters the machine and moves a cylinder? Entering the machine will de-energize the master circuit via the guard doors and dumps the air supply. An operator may need to move a cylinder to clear a jam and may forget to move it back. How the machine responds is your choice.

Let the cylinder move back. Letting the cylinder move may not be a problem when the operator is clear of the machine. Before the master can be energized, the guard doors need to be closed, thus keeping the operator away from the cylinder.

Add the *Fault Master Is Off* bit to each output latch, as shown in Figure 3-13, to prevent motion after motive power is restored. Now it is a recovery problem. Recovery options include aborting the sequence and starting over, backing the sequence up a step or two, or tell the operator to go back and make things right. Forward checking can be used to check the machine position and alert the operator when the logic and physical world are not synchronized. Forward checking is discussed in Chapter 5 on checking sensors.

3.9 Three-Position Valves

Three-position valves are similar to double-ended valves in that they have two solenoids. The difference is that the spool of the three-position valve

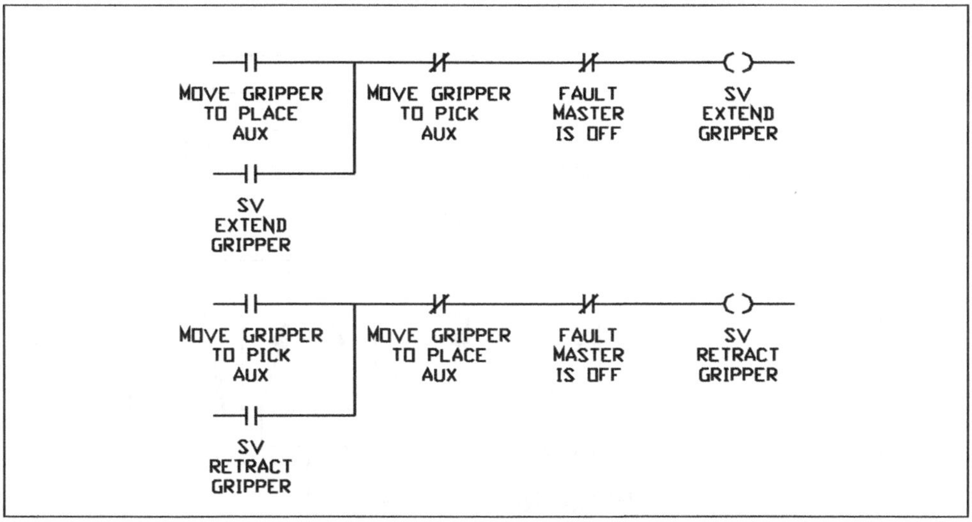

Figure 3-13. Using the Master Circuit to De-energize the Double-Ended Valve Outputs

will return to a center position when both solenoids are de-energized. The center position can be chosen to make the connected cylinder accomplish different tasks.

- **Open Center.** The center position can exhaust both sides of the cylinder to relieve any force the cylinder may have.
- **Blocked Center.** The center position can block whatever pressure is in both side of the cylinder to freeze its current position.
- **Pressurized Center.** The center position can pressurize both end of the cylinder to center it in the middle of its stroke.

The choice of the center position depends on the application. Figure 3-14 shows a three-position valve where the center position exhausts both sides of the cylinder.

The three-position valve is controlled with two output latches, as shown in Figure 3-12. This is the same logic as used with a double-ended valve. This logic, coupled with an open center valve, makes it easier for the operator to move the connected cylinder around when motive power is removed. A blocked center valve makes it difficult to position the cylinder by hand.

3.9.1 Relieving Pressure

One use of a three-position valve is to provide compliance. A cylinder can position a component part, then allow it to move around so that guide features on the tooling can locate the part more precisely. The sequence first extends the cylinder to present the component part to the tooling. Next,

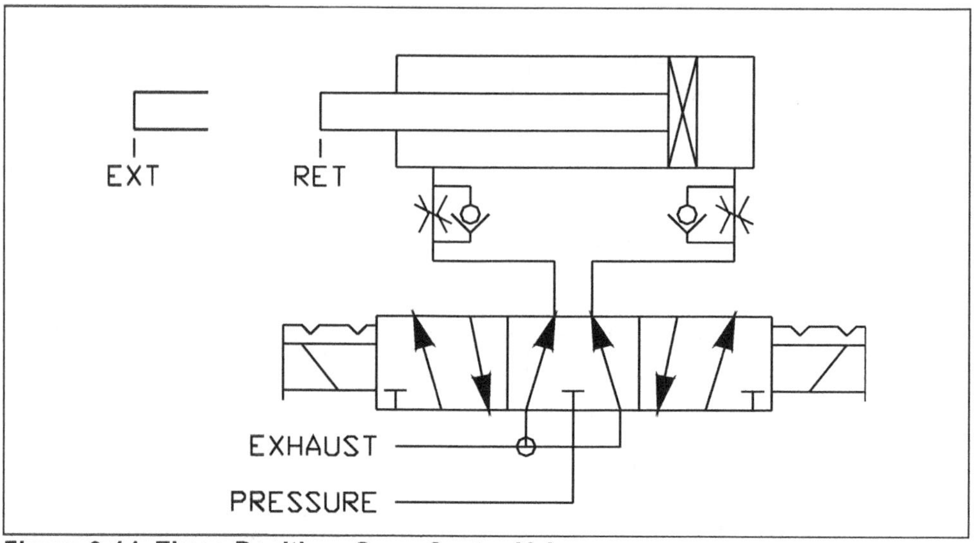

Figure 3-14. Three-Position, Open Center Valve

the pressure to the cylinder is relieved to let the part move. Finally, after the operation is finished, the cylinder is retracted to remove the part. This can be accomplished by using three auxiliary rungs and two output latches. The auxiliary rungs sum up the commands to extend, relieve pressure, and to retract. They control the output latches, as shown in Figure 3-15. Notice that the logic configuration has good push-pull action, in that only one solenoid can be energized at a time.

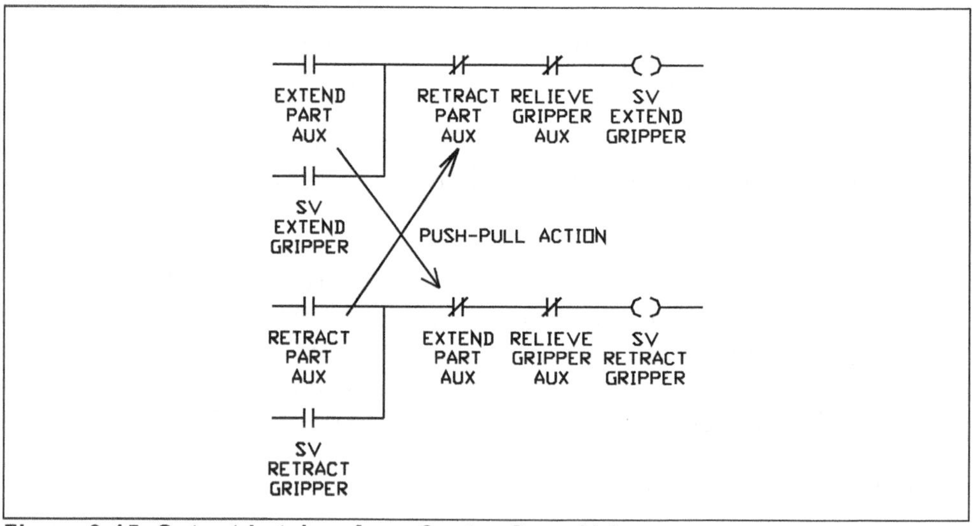

Figure 3-15. Output Latches for a Center Open Valve

3.10 Three-Position Cylinders

The last example to illustrate the versatility of the output logic involves a three-position cylinder. Suppose the pick/place mechanism of our example is modified to include sensors and logic to inspect the part before it is picked. The horizontal motion of the unit is also upgraded to a three-position cylinder, to add a reject position where nonconforming parts can be dropped. The gripper now has three positions to choose from: pick, reject, or place. One way of plumbing the cylinder is shown in Figure 3-16, using a double-ended valve and a spring-return valve for a total of three solenoids.

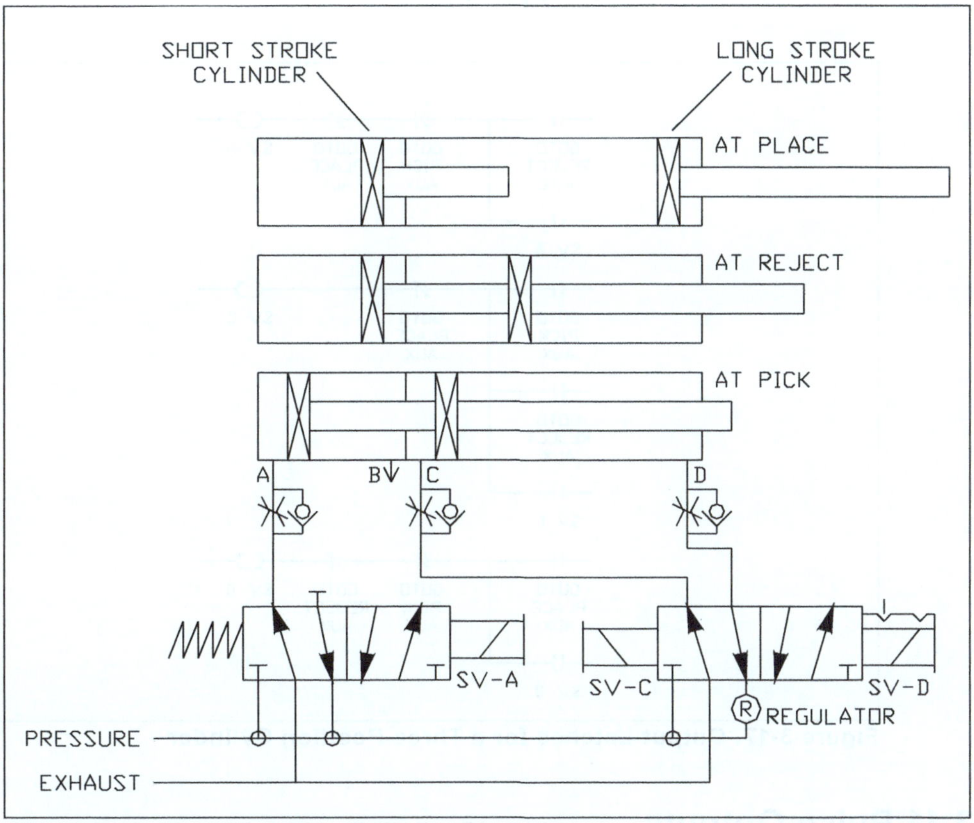

Figure 3-16. Three-Position Cylinder Plumbing

The output logic will consist of three auxiliary rungs to sum up the commands requesting either the pick, reject, or place position. Since there are three solenoids, there will be three output latches, as shown in Figure 3-17. Several notes just to be complete:

1. The cylinder port B is a muffler built into the cylinder. This is a common way of purchasing three-position cylinders.

2. The regulator is adjusted to a point below the line pressure. Without the regulator, the short-stroke cylinder will not develop enough force to push the other cylinder to the reject position. The regulator also has to be of the self-relieving type to accommodate the air pushed out of cylinder port D.

3. The short-stroke cylinder can only retract when the long-stroke cylinder retracts. This is because the short-stroke cylinder cannot be pressurized to make it retract; there is a muffler in cylinder port B. As a result, it does not matter if solenoid A is energized or not; when the gripper is at the place position, it is your choice. In this example, solenoid A is de-energized when the gripper is commanded to the place position.

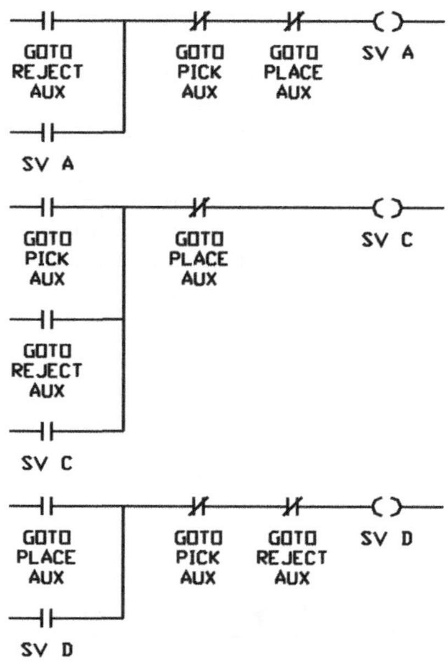

Figure 3-17. Output Latches for a Three-Position Cylinder

3.11 Pulse Outputs

The purpose of the motion object is to initiate motion and to monitor the motion. This implies the prime mover provides some type of feedback. Some devices do not provide feedback, such as simple impact printers or very short-stroke cylinders. The motion objects only tell these devices to operate and either assumes they worked or rely on other means to check the process. A downstream vision station can be used to check if the printer marked the component part or not.

A common method of starting simple devices is to provide a pulse that is on long enough for the device to see and start but not on long enough to make it operate over and over. Another example is a cylinder that only needs to extend long enough to bump a component part over to another conveyor lane. An example of a pulsed output is shown in Figure 3-18.

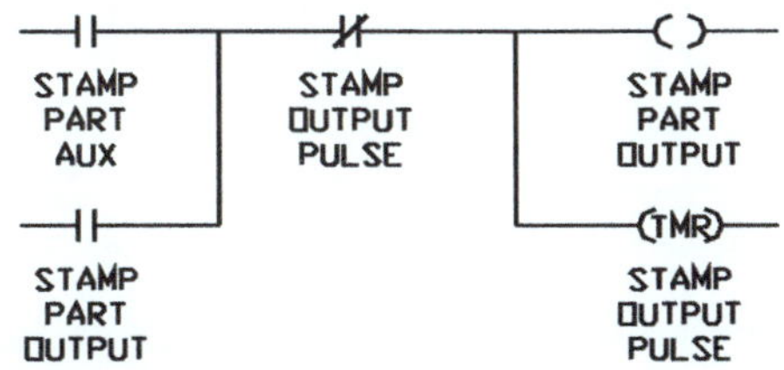

Figure 3-18. A Pulsed Output

3.12 Conclusion

As can be seen from the examples, the presented output logic is quite versatile. It is adaptable to many different types of prime movers and allows easy interfacing with the manual-mode layer. Development time is shorter because we have to remember only a few general logic structures. The end program is also more consistent and readable.

Gravity, and how to plan for it, is one of the biggest concerns with output logic. In general, we want motion to occur only when the machine logic tells it to, and preferably only after the operator presses the start button or initiates motion in manual mode. That is when everybody expects motion to occur. If motion occurs because of gravity, then keep it confined within the perimeter of the equipment's safety devices, such as behind guard doors and away from the operators.

4 Simulating Sensors

4.0 Introduction

The basic concept of a motion object is to initiate an action and to monitor the initiated action. Checking the action and determining if it finished or failed is important, not only for reporting errors but also to coordinate when to start the next action. The *Gripper Is Down* proximity switch on a pick/place mechanism makes sure the vertical motion is fully down before allowing the gripper to open or close; otherwise, the gripper may drop the part or grab it wrong. The feedback tells the logic when to start a task at the earliest possible moment. Using feedback to advance the machine's sequence is called *event-driven logic*. However, in many instances, time is used instead of a sensor, and this is called *time-driven logic*. Sensors may not be used for a variety of reasons, from reducing cost to the difficulty in applying them.

This chapter will show how time can be used in motion objects and in output logic to replace sensors. When time is used in a motion object it is called time-driven logic. When time is used in output logic its called *simulating sensors*. Both types can be used within the same sequence.

4.1 Cylinders with Only One Sensor

Suppose the vertical motion on a pick/place mechanism has a single proximity switch to tell the logic that the gripper is raised. The sensor indicating the gripper is lowered is not going to be installed. It would be difficult to install because the up/down stroke is so short. For instance, if the vertical travel distance is only 1/2 inch, which is about the diameter of a 12-mm proximity switch, fitting two large switches into the same area will be hard.

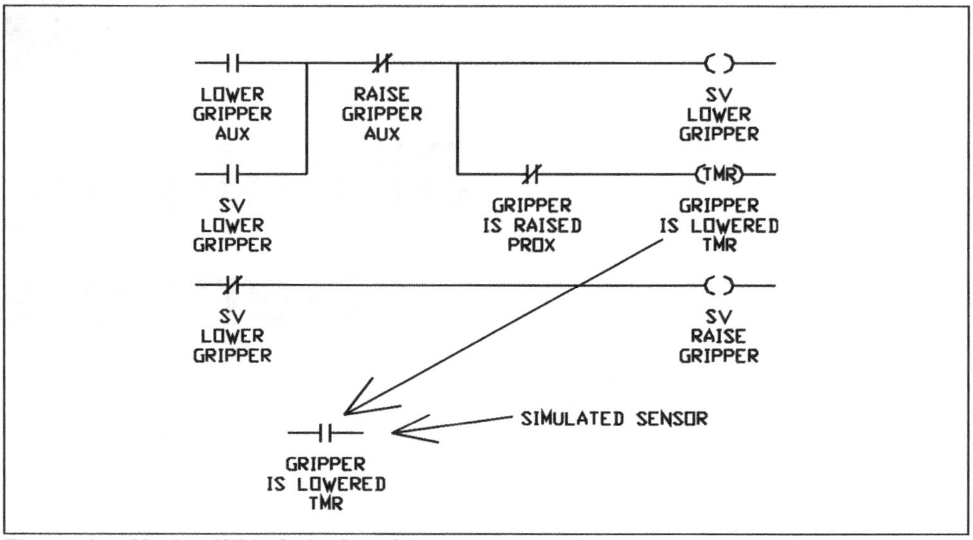

Figure 4-1. Motion With Only One Sensor

The output logic shown in Figure 4-1 uses a timer to create a *Gripper Is Lowered* sensor. This bit is used in the motion object in the same way an actual proximity switch is used. The timer's preset value is adjusted to match the actual travel time of the associated motion, in this case, the time it takes the gripper to lower. This simulated sensor uses as much knowledge of the real world as possible. Instead of using only time, it uses the facts that the grippers are commanded to lower and that the grippers are not raised. This is about as close to the real thing as you can get.

4.2 Cylinders with No Sensors

The output logic shown in Figure 4-2 uses two timers to create two sensors. The timers simulate the *Gripper Is Lowered* and the *Gripper Is Raised* sensors. The simulated sensors use the knowledge if the respective solenoids are fired or not.

4.3 Time-Driven Motion Objects

The standard motion object contains three rungs, as shown in Figure 4-3. The last rung is a fault latch that sets if the step does not finish within a certain amount of time. This rung is not needed when the step is time-driven. You can also get rid of the watch dog timer branch in the motion-initiate rung.

Figure 4-4 shows the simpler motion object used for a time-driven step.

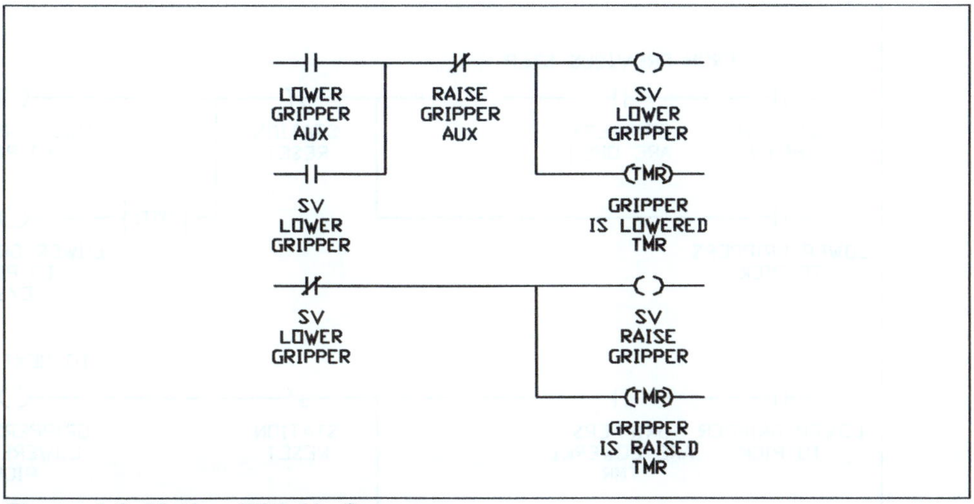

Figure 4-2. Motion With No Sensors

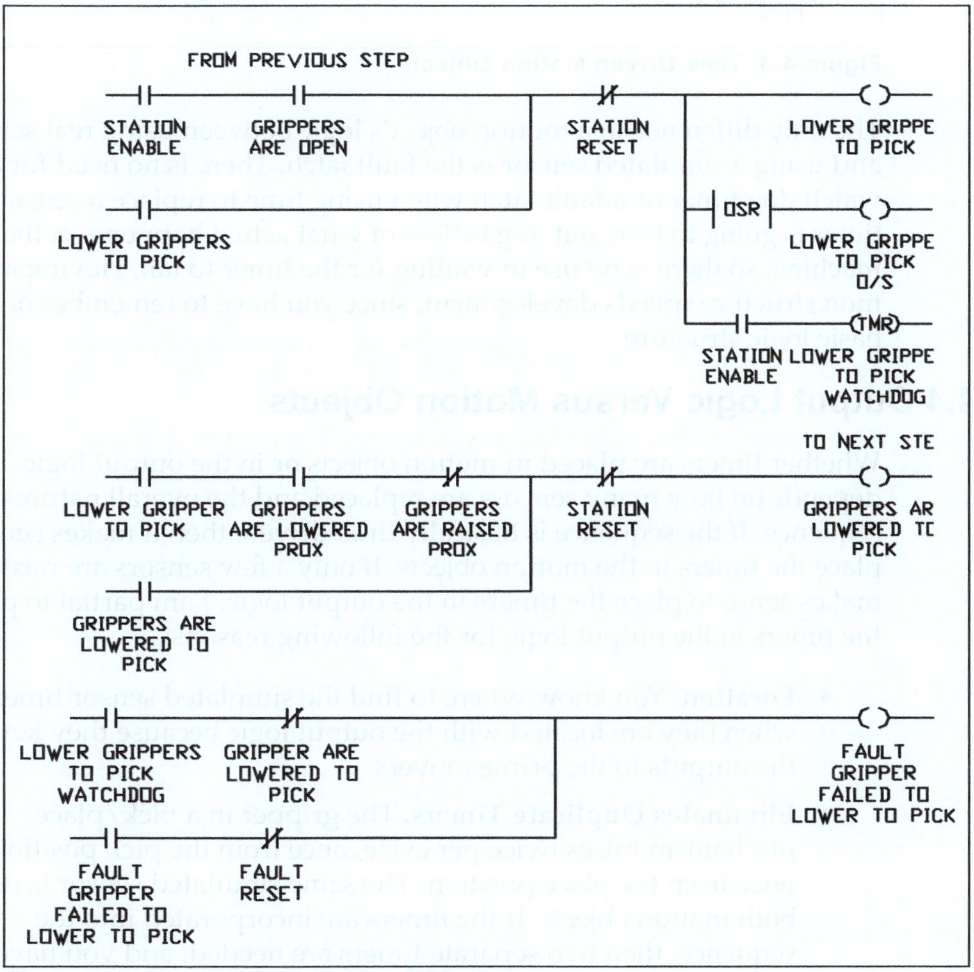

Figure 4-3. Event-Driven Motion Object

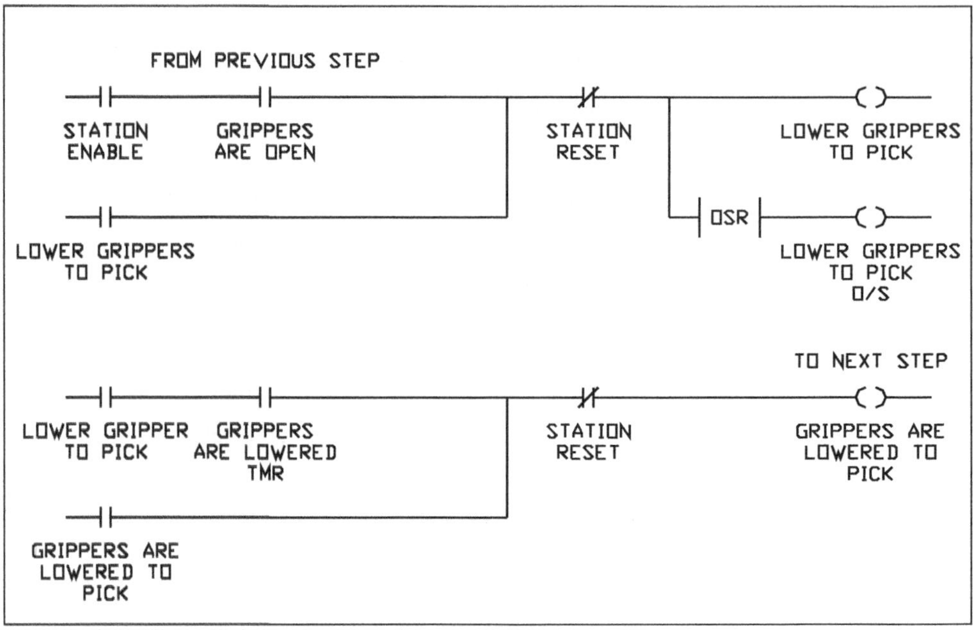

Figure 4-4. Time-Driven Motion Object

The only difference in a motion object's logic between using real sensors and using a simulated sensor is the fault latch. There is no need for a watch dog timer or a fault latch when using time to replace a sensor. The timer is going to time out, regardless of what actual happens on the machine, so there is no use in waiting for the timer to fail. Having a common structure speeds development, since you have to remember only one basic logic structure.

4.4 Output Logic Versus Motion Objects

Whether timers are placed in motion objects or in the output logic depends on how many sensors are replaced and the overall nature of the sequence. If the sequence is basically time-driven, then it makes sense to place the timers in the motion objects. If only a few sensors are missing, it makes sense to place the timers in the output logic. I am partial to placing the timers in the output logic for the following reasons:

- **Location.** You know where to find the simulated sensor timers when they are located with the output logic because they key off the outputs to the prime movers.
- **Eliminates Duplicate Timers.** The gripper in a pick/place mechanism raises twice per cycle, once from the pick position and once from the place position. The same simulated sensor is used in both motion objects. If the timers are incorporated into the sequence, then two separate timers are needed, and you have to remember to adjust both timers when a change occurs.

5 Checking Sensors

5.0 Introduction

This chapter discusses several common techniques used to detect and prevent problems caused by bad inputs, mainly sensors. Typical input problems include:

- Loose wires (every engineer's nightmare)
- Broken wires (fatigue from flexing is one of the biggest causes of trouble)
- Short circuits
- Overloaded or failed power supplies
- Incorrectly adjusted sensors (typical during startup)
- Electrical noise (welders, radios)
- Sensors failing on
- Sensors failing off

The inputs are the eyes and ears of the machine. False information can direct the machine to make wrong decisions, which can jeopardize the safety of the operator and lead to mechanisms crashing, loss of production, and loss of product. The machine logic has to protect the operator, protect the machine, and if possible, protect the product, generally in this order. Dangerous and explosive product will, of course, need to move up on the priority list. The machine, as a minimum, should alert and direct the operator to the input problem. Downtime is reduced by informing the operator where to start looking.

The machine can be protected using redundant inputs and the available sensors in combination. Redundancy uses two sensors to report the same condition in case one sensor should fail. Redundancy is typically not used and is reserved for critical situations. The common approach is to use the sensors you have in combinations to detect and prevent trouble. A pneumatic cylinder is generally equipped with two sensors: one to indicate when the cylinder is retracted and the other to report when the cylinder is extended. These two sensors provide all the information we need. The main thrust of the following techniques is to use the available sensors to detect trouble before it can cause damage.

Chapter 2 presented a motion object as the basic logic structure used to represent a single step of a sequence. The motion object uses a motion-complete rung to determine when the specific step is finished and to signal the start of the next step. It is in this rung where appropriate inputs are placed. Let's take a look at what could go wrong when a lone sensor is used.

5.1 Problems with Lone Sensors

To examine the potential problems with lone sensors, let's look at a typical pick/place mechanism. The sequence steps are listed in Figure 5-1 and illustrated in Figure 5-2. The sequence logic for Steps 7 and 8 is shown in Figure 5-3.

Step 1. Lower Grippers to Pick
Step 2. Close Grippers
Step 3. Raise Grippers from Pick
Step 4. Move Grippers to Place
Step 5. Lower Grippers to Place
Step 6. Open Grippers
Step 7. Raise Grippers from Place
Step 8. Move Grippers to Pick

Figure 5-1. Typical Pick/Place Sequence

5.1.1 Lone Sensor Fails Off

Consider what happens when the *Grippers Are Raised Prox* input (the only source of feedback) fails. If the input fails in the "off" state, the motion-complete latch, *Grippers Are Raised From Place,* will never turn on. The sequence then waits until the watch dog timer expires and sets the *Fault Grippers Failed To Raise From Place* latch. The sequence is then disabled

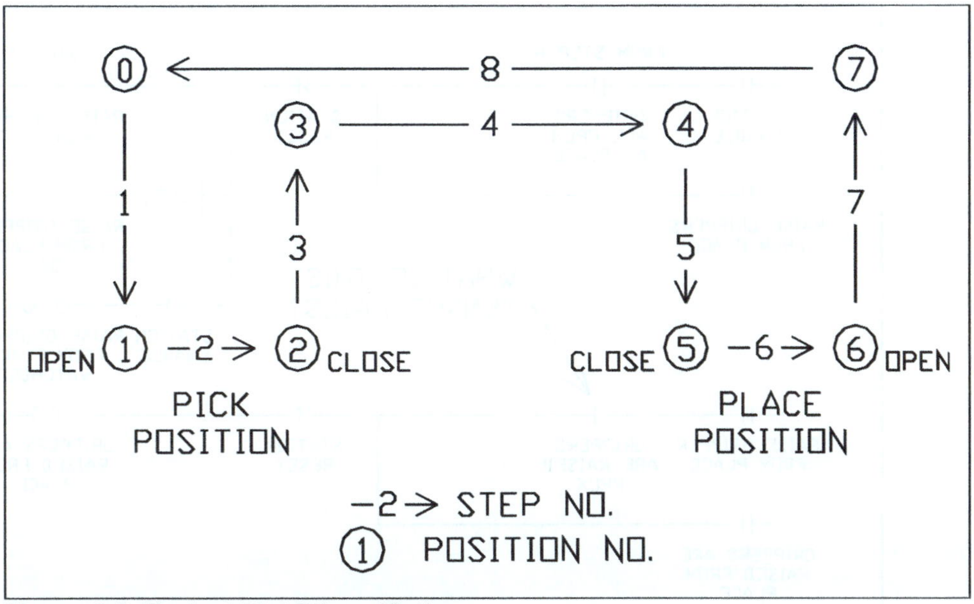

Figure 5-2. Typical Pick/Place Motions

from continuing by having the fault drop the *Station Enable* bit. The fault also alerts the operator to the failure via the message control logic. The logic for displaying messages is discussed in Chapter 13 on message logic.

5.1.2 Lone Sensor Fails On

Now consider what happens when the *Grippers Are Raised Prox* input fails in the "on" state. Immediately after the grippers are commanded to raise (Step 7), it is falsely reported in the raised position. The motion-complete rung, *Grippers Are Raised From Place,* will immediately turn on, which in turn immediately commands the grippers to move horizontally (Step 8). As a result, the grippers will start to raise and move horizontal at the same time. This will most likely result in a crash, as shown in Figure 5-4. One way of preventing this is to use a concept called *back checking*.

5.2 Back Checking

Back checking is using both sensors of a motion in pairs. Figure 5-5 is the motion-complete rung of Step 7, modified to include back checking. To report a cylinder is raised, the sensors have to indicate it is raised and not lowered. As can be seen, back checking is a simple modification. When we examine the entire pick/place sequence using only lone sensors, we find that the *Grippers Are Raised* sensor is used twice, in Steps 3 and 7, where it is actually needed to tell if the initiated action completed. If the sensor fails in the "on" state, the pick/place unit may crash. By adding back checking, the sensor, *Grippers Are Raised*, will also be used in Steps 1 and 5. The sen-

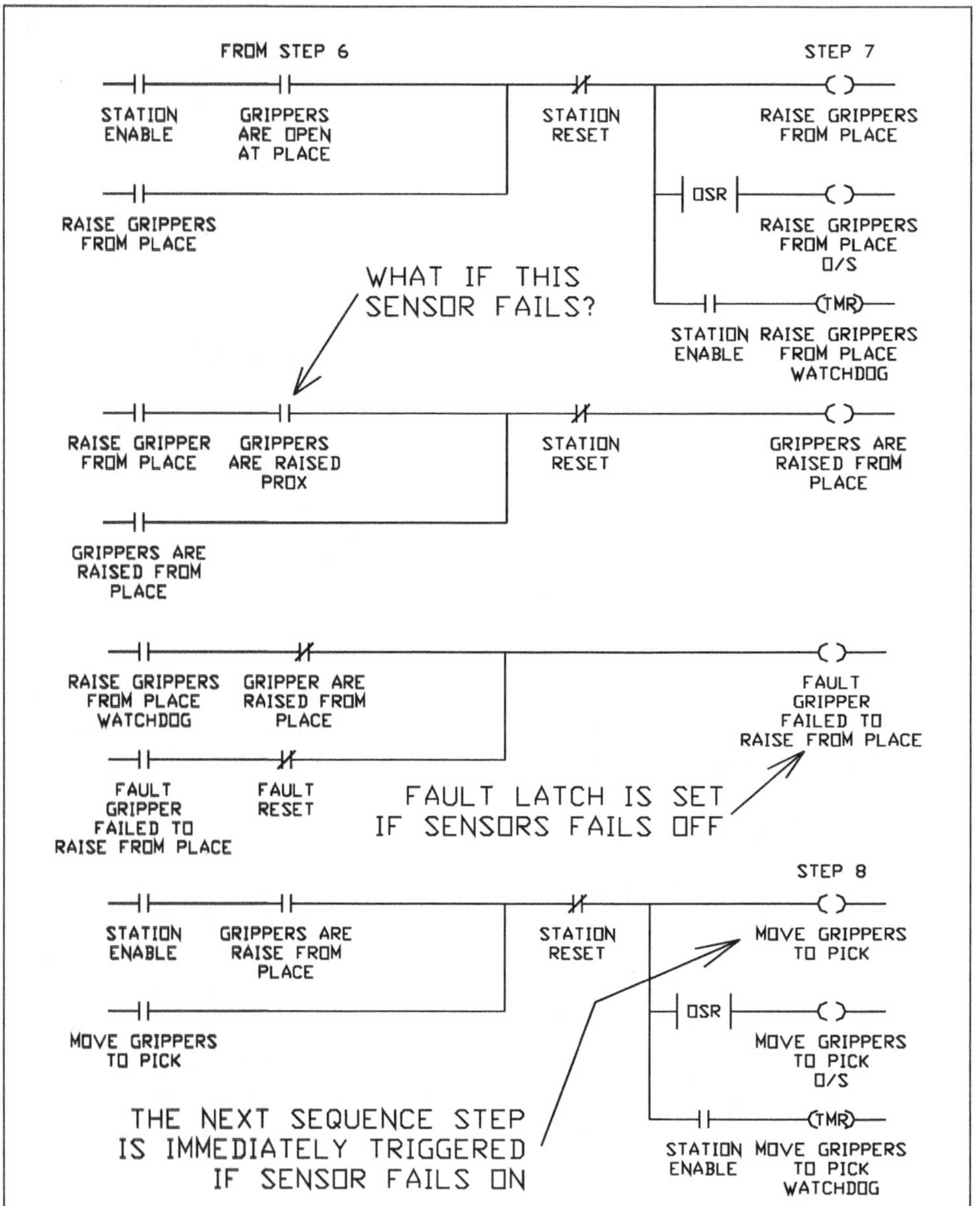

Figure 5-3. Pick/Place Logic for Steps 7 and 8

sor is now used twice as often, for a total of four times. The motion-complete rung for Step 5 is shown in Figure 5-6. In Steps 1 and 5, the failed *Grippers Are Raised* sensor will fault the step instead of causing any damage. The *Grippers Are Raised* sensor is not needed in Steps 1 and 5 to tell us the grippers are lowered, but it gives the machine logic a chance to see if the sensor failed. The machine logic now has two chances of catching the failed sensor.

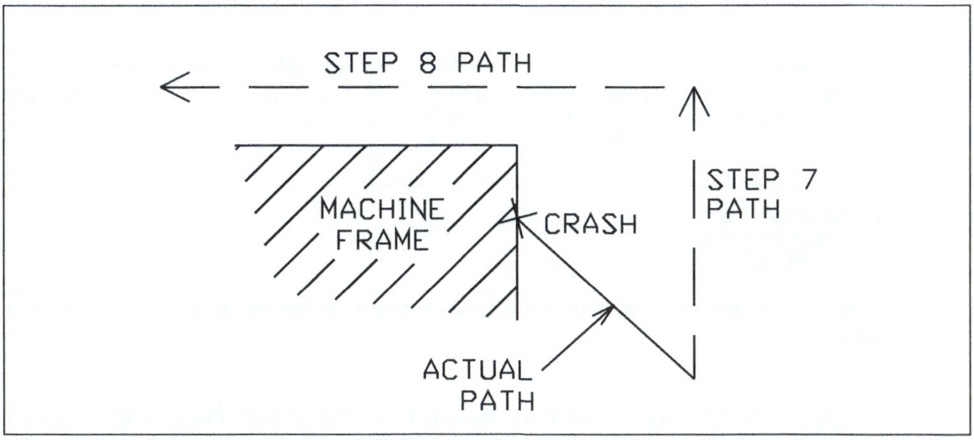

Figure 5-4. Consequences of *Grippers Are Raised* Input Failing On

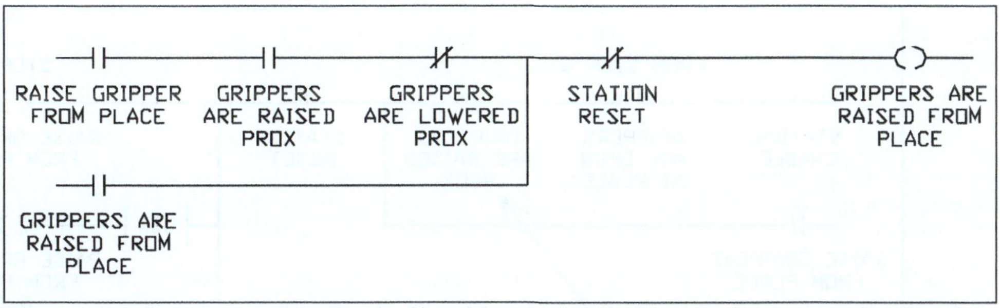

Figure 5-5. Incorporating Back Checking into the Motion-Complete Rung of Step 7

Mechanism may still crash with back checking, but the probability of a sensor failing between Steps 1 and 3 or between Steps 5 and 7 is low. Back checking does give the machine logic a chance of detecting a failed sensor, where there was no hope at all without it. Back checking is required by many buyers of capital equipment and is considered throughout the industry as a sufficient means by itself to detect and prevent problems. However, back checking cannot guarantee the prevention of a crash. To do this we need forward checking.

5.3 Forward Checking

Forward checking works similarly to back checking, by looking at a sensor before it is needed. Back checking does this by looking at the *Grippers Are Raised* sensor in Steps 1 and 3, where the grippers are commanded to lower. Forward checking expands on this by looking at the appropriate sensor immediately before it is needed to eliminate the blind spots inherent in back checking. Figure 5-7 shows Step 7 of the pick/place logic modified to accept forward checking. The modifications include adding the

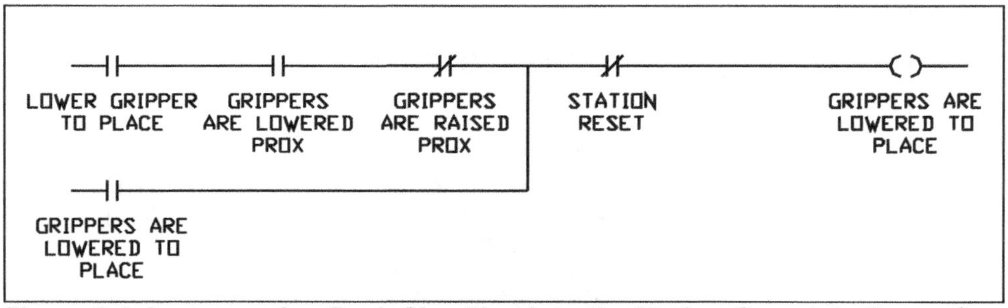

Figure 5-6. Incorporating Back Checking into the Motion-Complete Rung of Step 5

Grippers Are Raised sensor to the motion-initiation rung and a new fault latch.

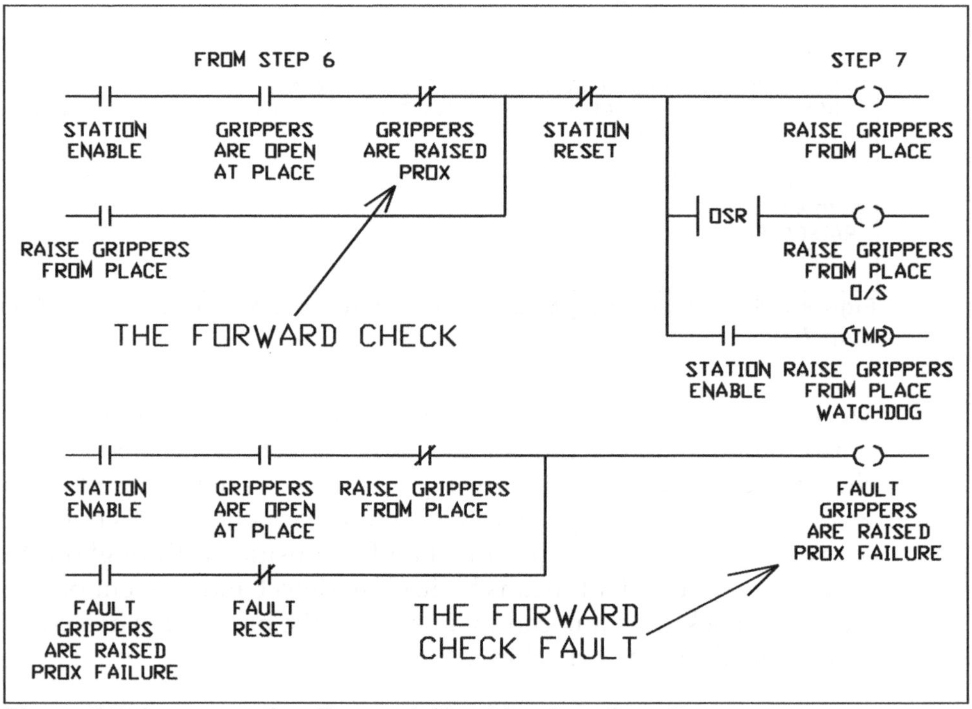

Figure 5-7. Incorporating Forward Checking into Step 7

The *Grippers Are Raised* sensor in the motion-initiation rung will prevent the grippers from raising if the sensor is already reporting the grippers are raised (sensor failed on). This will also keep the next sequence step from triggering to prevent a crash. The new fault rung, *Fault Grippers Are Raised Prox Failure,* was added to alert the operator. Without this fault, the sequence will stall, and the operator will not know why the machine stopped. This fault is called the *forward check fault* and will inform the operator if the sensor fails.

Forward checking will prevent crashes caused by faulty inputs, but it does cost extra programming labor. Using forward checking only at critical sequence steps can save some effort. Consider the following when deciding to use forward checking:

- Can a mechanism crash damage a valuable product?
- Can a mechanism crash result in excessive downtime?
- Can damaged product release toxic agents?
- Can damaged product blow up?
- Is there an increased probability of sensor failure because of excessive vibration or chemical exposure?

5.4 Position Bits

One advantage of using a common programming template, such as a motion object, is to ease the task of programming. In theory, once the flow chart is created (or some type of game plan is formed), it is a simple matter of generating code by replacing the sequence steps with motion objects. Using a mixture of back checking and forward checking would defeat the common template approach and thus increase our efforts. On the other hand, using forward checking on every step just to be consistent would again increase of efforts. There is a compromise called position bits, which Vern Werth of Remmele Engineering, Inc located in St. Paul, Minnesota developed. It allows a common template and forward checking with only a slight increase in effort.

Position bits maximize the probability of catching a failed input by including it twice in every motion object of the sequence: once at the forward check in the motion-initiation rung, and once in the motion-complete rung. Instead of inserting many normally open and normally closed contacts at each location, an appropriate position bit is inserted. A position bit is a summation of all of the relevant inputs in such a combination to describe a physical position the mechanism can be moved to. Some position bits for a pick/place mechanism are shown in Figure 5-8.

Figure 5-9 illustrates Step 7 of the pick/place mechanism using position bits. As you can see, the position bits really clean up the logic. The labor of incorporating forward checking is significantly reduced, with most of the added labor going to the forward check fault.

5.4.1 Fault-Recovery Option

Imagine your sequence is frozen by some fault knocking out the *Station Enable* bit. The operator can switch to manual mode and start command-

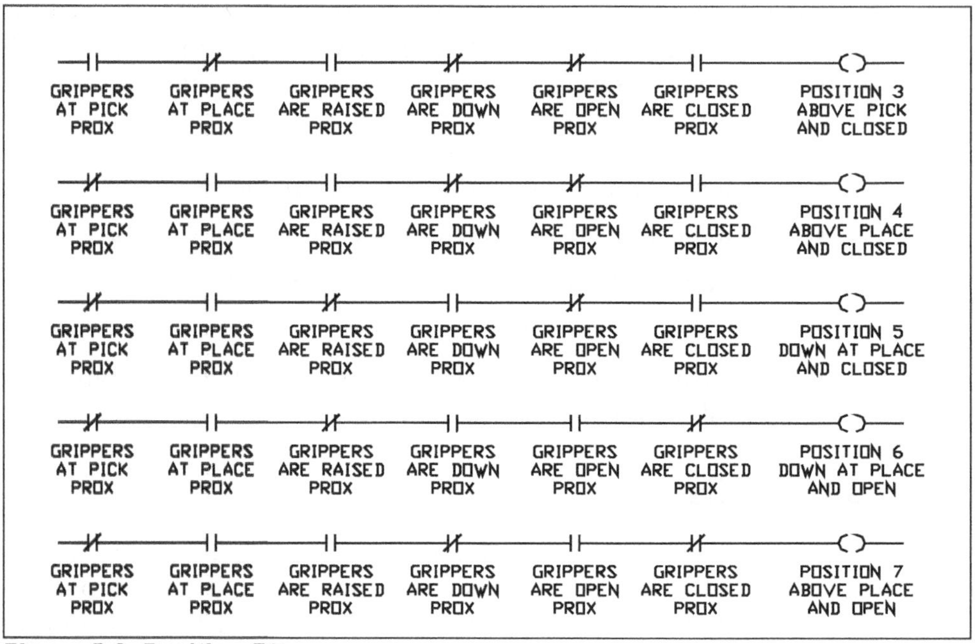

Figure 5-8. Position Bits

ing cylinders to move all over, doing whatever is needed to deal with the fault. The operator can even enter the machine and move mechanisms by hand. The operator then switches back to automatic mode and restarts the machine. If the operator placed all the moved motions back to their original positions, the machine will continue where it left off. If just one motion was inadvertently left in the wrong position, the forward check fault will catch it. The forward check faults will tell the operator which specific motion was misplaced.

The ability to interrupt the automatic cycle and allow manual operations provides another method of fault recovery. It also allows easier machine setup and is a tremendous help during debugging. The operator still has to return motions to previous positions, but now the machine is protected and the operator is informed of any errors.

5.5 How Often Sensors Are Checked

Figure 5-10 counts the number of times the *Grippers Are Open* input is used for each type of sensor-checking scheme. The count was determined from Steps 5 through 8 of the pick/place sequence. As seen in Figure 5-10, the sensor occurrence skyrockets when using position bits.

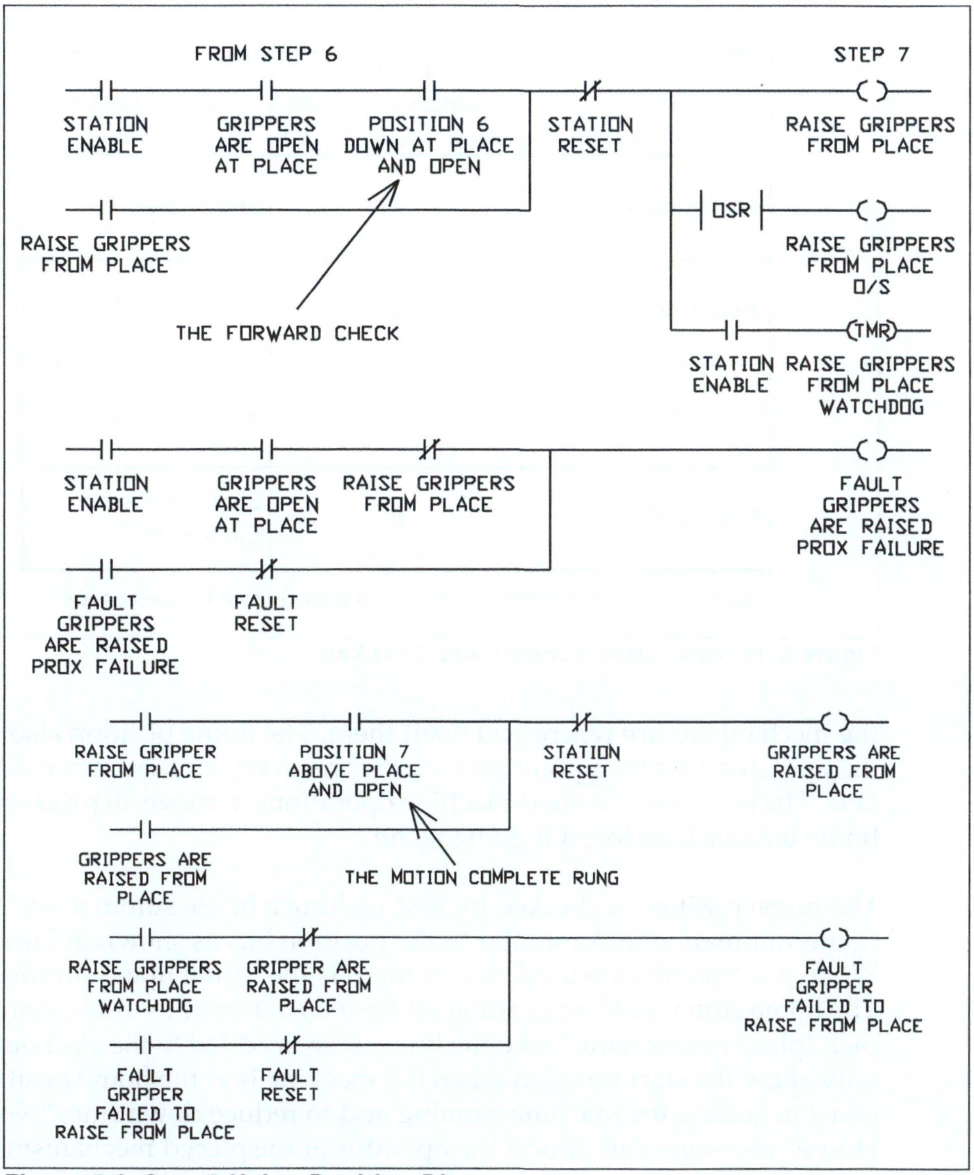

Figure 5-9. Step 7 Using Position Bits

5.6 Home Position

There is another occurrence in the machine logic where sensors are checked. This check ensures all the mechanisms are at a specific, well-defined "home" position before the machine can be started. The home position ensures that all of the mechanisms are synchronized with the controlling logic. When a sequence starts executing, it assumes the mechanisms are in some initial position then commands them to move from there. Requiring the machine to start from the home position guarantees

Sensor Configuration	Quantity *Grippers Are Open* Sensor is Checked	Where Sensor is Checked
Single Sensor	1	End of Step 7
Back Checking	2	End of Steps 5 and 7
Forward Checking[1]	3	End of Step 5 and Start and End of Step 7
Position Bits	8	Start and End of Each Step

[1]Back-checking at completion of move and single sensor forw ard-check

Figure 5-10. How Often Sensors Are Checked

the mechanisms are where you want them. The home position also gives the operator a starting point and another recovery method when all else fails. The operator can abort machine operations, remove all products, and home the machine to get it going again.

The home position is checked by first making a home summation. The home summation looks similar to the position bits as shown in Figure 5-8. When position bits are used, one of them is the actual home summation. The home sum would be position bit *Position 0 Above Pick And Open* for the pick/place mechanism. Next, the home sum is added to the start button to only allow the start function when the machine is at the home position. To assist in getting the machine running and to reduce downtime, "Not Home" messages can inform the operator of misplaced mechanisms.

Starting is different than resuming. When automatic operations are resumed, the mechanisms continue from where they left off. In this case, the machine is not required to be in the home position. Chapter 6 on controlling sequences explains how the logic can differentiate between starting and resuming. Maintaining synchronization between the position of mechanisms and the sequence logic can be a problem when resuming operations. It is possible to move mechanisms around by hand, when a sequence is holding cycle. Any moved mechanism needs to be returned to its original position. Two basic methods are used to guarantee synchronization. One method is to use forward checking as discussed above. The other method is more common and requires the operator to abort the sequence and "home" the machine. Homing buttons can be included with the operator interface to make this process more convenient and faster, but it is still slower than forward checking.

Figure 5-11 shows how the home check works with the other sensor checking schemes. Because of the cyclic nature of machinery, when a sequence triggers a cylinder to extend, later it will trigger the same cylinder to retract.

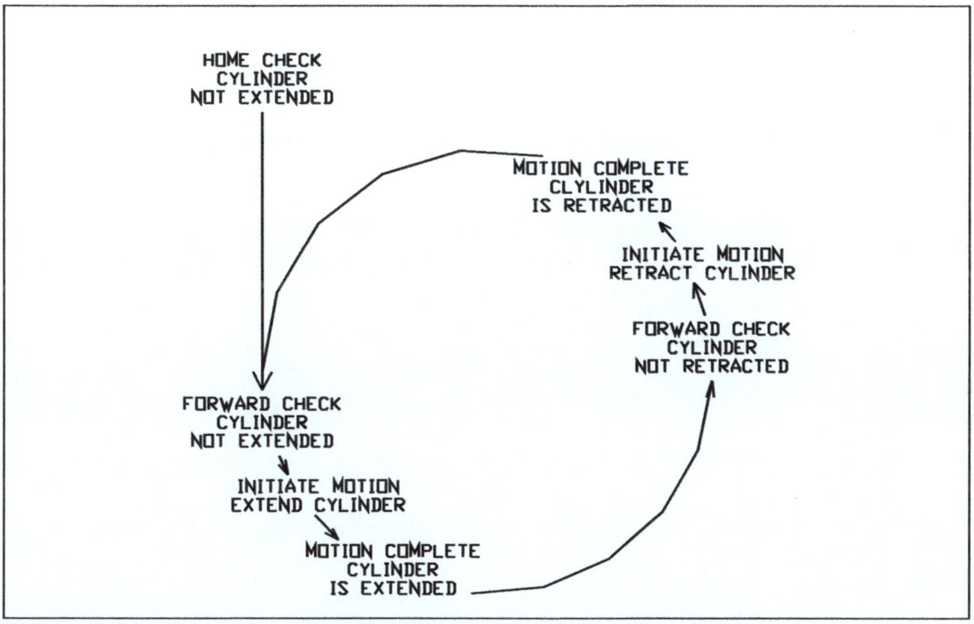

Figure 5-11. Where Sensors Are Checked In A Machine Cycle

6 Controlling Sequences

6.0 Introduction

At this point, we have sequences built from motion objects, motion objects linked to outputs, and reliable feedback. This chapter will discuss how to control the sequences by added logic to start, stop, freeze, resume, and abort a sequence. Other features of sequence control, such as stepping and single cycling, are discussed in the next chapter.

6.1 Types of Sequences

Sequences can be categorized on the basis of who controls them and how they interact with other sequences, as a parent, child, independent, and dependent sequence. No matter the type, they all share a common structure using several key latches to govern their control.

6.1.1 Independent Sequences

An independent sequence starts when the system control starts and repeats as long as the system control stays on. System control is discussed in Chapter 8. In a nutshell, the system control starts and stops the whole machine and controls the devices common to all the sequences. Independent sequences are used to control groups of mechanisms that are independent from the other mechanisms on the machine. There is typically no physical interference between the mechanisms of an independent sequence and the other machine mechanisms. A vibratory feeder bowl, with a part sorting mechanism that sends parts into a machine on a conveyor belt, is a good example. The bowl and sorter operate to keep product at the end of the conveyor and does not really care about the mechanisms taking parts off the conveyor. The conveyor provides the buffering of parts and eliminates the need for the bowl and sorter to interact directly with the rest of the equipment. The mechanism receiving the product only cares if product is present before processing it and only complains when the product fails to show up.

6.1.2 Dependent Sequences

A dependent sequence is a sequence that works with another sequence to perform a task. The two sequences are generally independent from each other, except when they need to interact. The sequences for dependent mechanisms operate as long as the system control is on, pausing only to interact with another sequence. A vibratory feeder bowl with a part sorter and a pick/place mechanism that transfer parts to a station on a dial is a good example. The bowl, sorter, and pick/place mechanism operate to present a part. Then it interacts with the station's mechanisms, when the station asks for a part. The bowl, sorter, and pick/place mechanism is programmed with its own sequence separate from the station's sequence. This allows it to find and present a part while the station is busy or unable to operate, such as when the dial is indexing. Parent/child sequences are dependent sequences but are given their own classification due to the unique interlocking requirements between these types of sequences.

6.1.3 Parent Sequences

A parent sequence controls other, child, sequences. Parent sequences control the mechanisms in common to other groups of mechanisms on the machine. They start when the system control starts and repeat as long as the system control stays on. The parent sequence is used to synchronize its children and the data that flows between them. The sequences controlling a dial or a walking beam are good examples because when these mechanisms finish transferring parts between the stations, they command the stations to operate.

6.1.4 Child Sequences

A child sequence starts when commanded by a parent sequence and only operate once for each command from the parent sequence. Stations on a dial machine are controlled by a child sequence, with the dial being the parent sequence. The dial sequence rotates the dial and then tells each station to operate.

6.2 Common Control Structure

All the different types of sequence share six common rungs with each rung serving a unique purpose. The six key rungs are:

- **Station Home.** Mechanisms are at their starting positions
- **Station Enable.** Controls sequence execution, run or hold
- **Run Station.** Starts a sequence and remembers it was started
- **Station Reset.** Resets all the sequence latches so the sequence can repeat
- **Station Fault Sum.** A problem within the sequence
- **Station Running.** Sequence is executing and causing motion

A typical sequence will have the control rungs, sequence steps, and output logic placed in the order shown in Figure 6-1. The shift register logic will be discussed in more detail in Chapter 10.

Station Home
Station Enable
Run Station
Inspect Shift Register
First Motion Object
Other Motion Objects
Last Motion Object
Update Shift Register
Station Reset
Station Fault Sum
Station Running
Station Output Logic

Figure 6-1. Typical Sequence Logic Order

6.2.1 Run Station

The *Run Station* latch serves two purposes: It controls when the sequence is started and remembers that the sequence was started. A typical *Run Station* latch is shown in Figure 6-2.

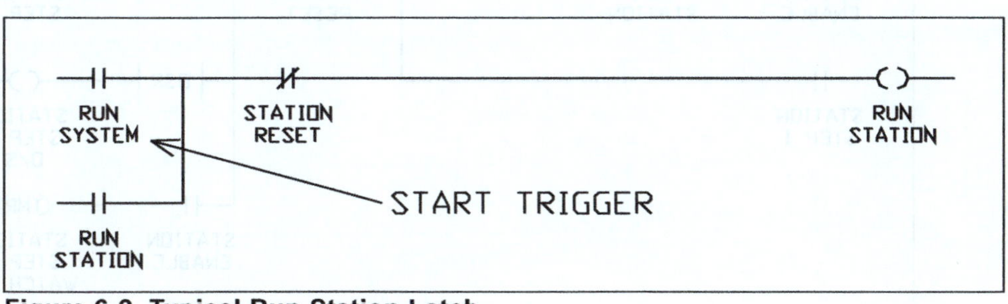

Figure 6-2. Typical Run Station Latch

Triggering the *Run Station* latch starts a sequence. The trigger originates from logic external of the sequence, and it can come from another sequence or directly from the system control. The trigger can be a pulse (one shot) or a bit that remains on for as long as the machine is on.

The *Run Station* latch triggers the first motion object in the sequence, which in turn triggers the next motion object and so on, until the last motion object triggers the *Station Reset* bit. The *Station Reset* bit indicates the sequence is done and unlatches all the motion objects, any other sequence latches, and the *Run Station* latch. The sequence can immediately restart after it is reset if the start trigger remains on; otherwise, the sequence will wait for the next trigger. Notice that turning the start trigger off does not stop the sequence mid-stream but only prevents it from restarting, thus defining a cycle stop point. Cycle stopping a machine is useful because it lets stations finish their tasks, instead of letting product dangle in mid-air or having incomplete operations.

Figure 6-3 shows the motion-initiation rung of the first motion object in the sequence. The start trigger could just as well be placed in this rung to start the sequence, to save writing an additional rung. Using a separate rung, such as *Run Station* serves to differentiate the functionality of each rung. The *Run Station* latch is used by the system control to determine if a sequence is starting fresh or will resume, and the first motion object is used to initiate a motion. Separate rungs allow separate coil comments to make the logic more readable, and it is easier to modify the sequence. A change within the sequence should be just that and not a change concerning the whole machine. Lastly, there is usually logic placed immediately after the *Run Station* latch to inspect shift register information, thus making separate rungs mandatory to time the inspection with the motions.

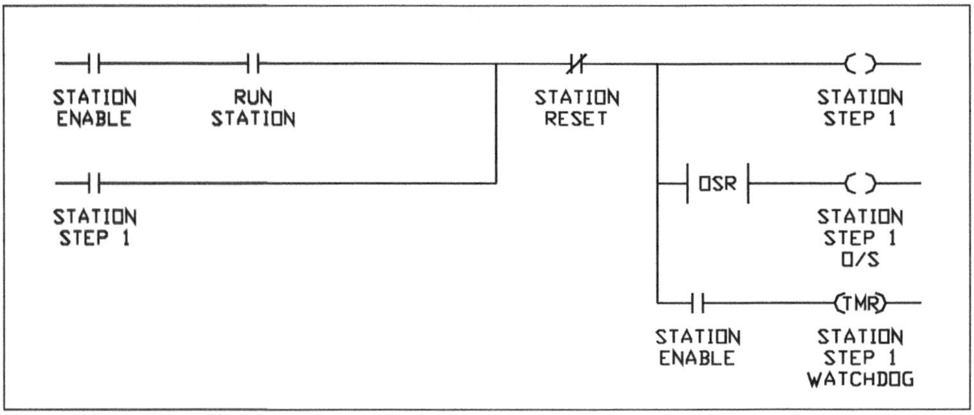

Figure 6-3. First Motion Object In A Sequence

In Figure 6-3, there is also a *Station Enable* bit included with the *Run Station* bit. The *Station Enable* bit basically controls freezing and thawing of a sequence. It is not needed in the first motion object, when the first motion object is immediately triggered by *Run Station*. The *Station Enable* bit is shown to indicate its location within each successive motion object.

After *Run Station* is triggered, it will remain on until the sequence is finished as controlled by the *Station Reset* bit. The sequence may not be currently executing because the *Station Enable* bit is off, but *Run Station* remembers that the sequence was started. As a result, we can examine *Run Station* to tell if the sequence will either wait for the next start trigger or resume execution from where it stopped. Knowing if a sequence will remain idle or will resume is important to the system control and the operator, as will be seen.

6.2.2 Station Enable

The *Station Enable* latch controls the execution of the sequence. It is used to stop a sequence between steps to allow stepping through a sequence and to pause it if trouble develops. The *Station Enable* bit is inserted in the motion initiation rung of each motion object, as shown in Figure 6-3. When the *Station Enable* latch is on, the sequence can step from one motion object to the next. When off, the sequences cannot trigger the next motion object, thus freezing the sequence in place. Notice that a motion object, once started, is allowed to finish. This defines a machine stop versus a cycle stop. Figure 6-4 shows a typical *Station Enable* latch.

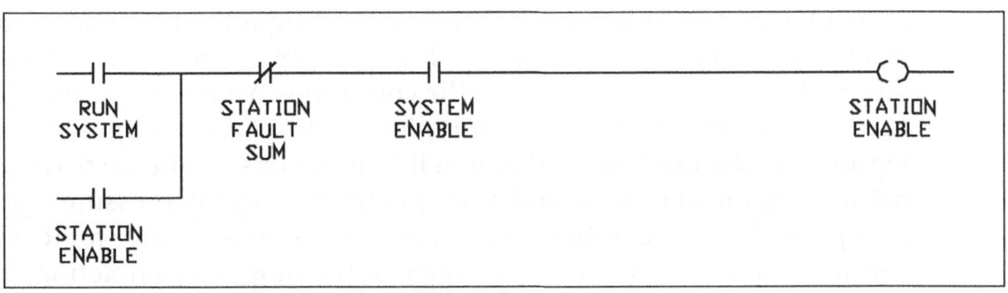

Figure 6-4. Typical Station Enable Latch

The *Station Enable* latch is triggered by the system control as shown by the *Run System* bit in Figure 6-4. The *Run System* latch turns on when the operator presses the cycle start push button provided the machine is safe, ready, and capable of running. When the machine is started for the first time, the *Station Enable* latch is turned on, but the sequence won't execute until the *Run Station* latch is triggered. The examples in Figures 6-2 and 6-4 show the *Run System* latch triggering both the *Station Enable* and *Run Station* latches. This is typical for independent sequences. Other sequence types are enabled when the machine is started but triggered to execute later in the machine cycle. When the machine is restarted, say after a fault stopped it, the *Run System* latch will re-trigger the *Station Enable* latch to allow an incomplete sequence to finish. The use of *Run System* lets only the operator, through the *Cycle Start* push button, to enable a sequence. An enabled sequence is more dangerous than a disabled one because an accidentally triggered sensor in an enabled sequence can cause a sequence step to function and create motion. The use of two latches, *Enable Station* and *Run Station*, lets a sequence be enabled when its safe to do so, and lets it run when it is needed to do so.

The *Station Enable* latch, once on, will remain on until there is a problem with the sequence (*Station Fault Sum*) or a problem with the machine as a whole (*System Enable*). Fault latches within the sequence are used to detect any problems with the sequence. Any condition that will prevent a sequence from finishing should be monitored, and a fault latch should be used to catch the condition. Fault latches can capture momentary conditions to let you know they did indeed occur, so you do not have to guess. Fault latches also alert the operator to problems, reducing downtime. Most of the fault latches are those incorporated within the sequence's motion objects. All the fault latches within a sequence can be summed up into one bit called the *Station Fault Sum*, to make using this information easier. The *Station Fault Sum* can then drop the *Station Enable* latch, to freeze the sequence.

The *System Enable* bit shown in Figure 6-4 is used to turn off the *Station Enable* latch. The *System Enable* bit is part of the system control. It is used to prevent and/or stop a sequence from executing if the machine is not safe or incapable of generating motion. It can disable a sequence at any time. The *System Enable* bit looks at all the conditions necessary for the machine to operate properly and safely. A typical *System Enable* rung is shown in Figure 6-5. The machine logic can tell if its actions can be carried out by monitoring the master control relay, pneumatic supply pressure, hydraulic pumps, and so on. Disabling a sequence prevents sequence faults by preventing motion when there is no hope of the motions completing. It also helps protect the operator from unwanted motions when the operator is inside the machine. Remember that once a motion object is started, it is allowed to finish. There is usually enough residual energy to let a motion finish once started when the *System Enable* turns off immediately after a motion object is started, but there is no guarantee.

Notice that a sequence will not be disabled if there is a problem with another sequence. Logic can be added to disable some or all the sequences if there is a problem with another sequence, but generally it is not done. Stations on a machine are usually segregated enough to allow one station

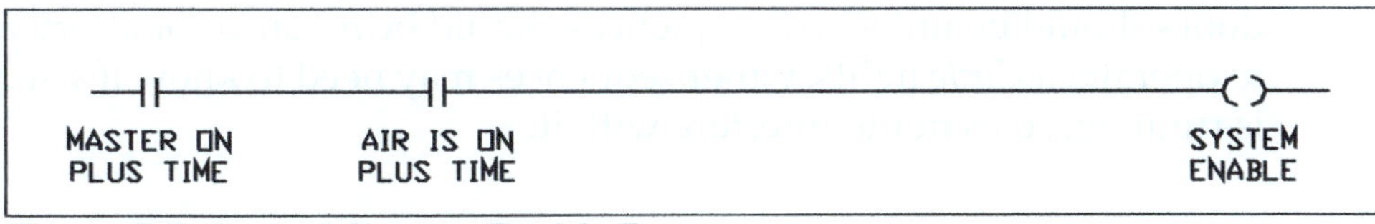

Figure 6-5. Typical System Enable Rung

to operate to a certain point without the other stations operating. A station cannot continue when it needs the product from an upstream station. One cycle of a station (one complete execution of its sequence) typically defines what the station needs to do to the product it currently has before it needs new product. As a result, when one sequence faults, there is no reason why the other sequences cannot finish. Letting the nonfaulted sequences finish is preferable over disabling (freezing) them. It puts the remaining stations in more stable conditions with completed product. A disabled station may hold a component part in mid-air instead of placing it securely in a fixture.

A *System Fault* sum is used to collect all the station fault summations and other system faults into one convenient bit. This bit is used to turn off the *Run System* latch. Without *Run System*, the sequences, once finished, will not be restarted. The overall result of individual *Station Enable* latches for each sequence is to let sequences finish but not to restart when another sequence faults and to disable and freeze the faulted sequence.

Figure 6-6 is a truth table illustrating how the *Station Enable* and *Run Station* latches can be used in combination external to the sequence. They tell the system control and other sequences if it is executing (running), or frozen, and so forth.

Run Station	Station Enable	State of Sequence
Off	Off	Sequence Off and Not Ready
Off	On	Sequence Off and Ready
On	Off	Sequence Frozen
On	On	Sequence Runing

Figure 6-6. Sequence States

6.2.3 Station Running

A *Station Running* bit is created to make it easier to know if a sequence is executing or not. This bit is shown in Figure 6-7. Knowing if a sequence is executing and possibly causing motion is useful when deciding if guard doors should be unlocked, sequences should be resumed or aborted, and to operate *In Cycle* lights. Other sequences may need to know if a sequence is running, so as not to interfere with it.

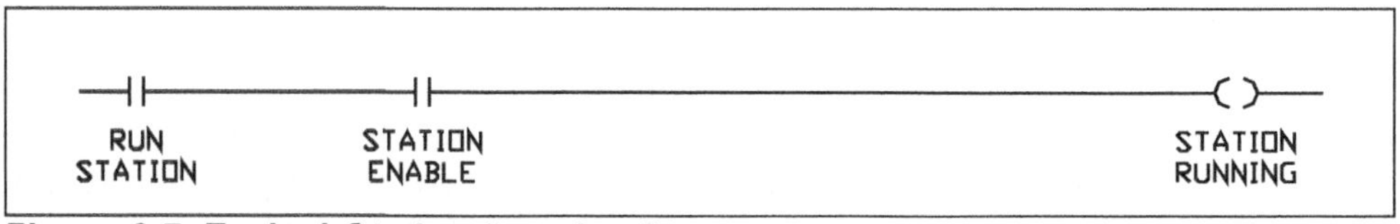

Figure 6-7. Typical *Station Running* Bit

6.2.4 Station Home

The home position, discussed in Chapter 5, is the position all the mechanisms need to be in when a sequence starts. A sequence proceeds by commanding its mechanisms to move from the home position, so it is important that the mechanisms be in this position when the sequence starts. Requiring the mechanisms to be in their home position before letting the associated sequence start guarantees the mechanisms and the sequence are synchronized. The guarantee is made by inhibiting the machine's start function (*Run System*) unless the mechanisms are in their home position. To help organize the logic, each sequence has a *Station Home* summation. An example of a home summation for a pick/place station is shown in Figure 6-8. All the individual home summations can be summed up into a *System Ready* summation for use by the system control.

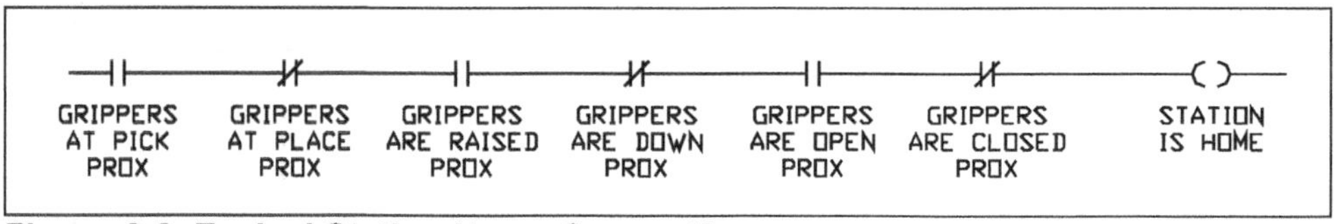

Figure 6-8. Typical Station Home Summation

A frozen sequence may not be in its home position. As a result, the system control will not be allowed to restart the machine unless the logic allows for this. The system control does know if a sequence is starting anew or resuming from its current position. The *Run Station* latch provides this information and can be used to condition the respective *Station Home* summation to allow the machine to start. This is shown in Figure 6-9. Bypassing the *Station Home* summation poses no problems, since a started sequence is already synchronized with its mechanisms. If there is a chance the mechanisms of a frozen sequence will move either by gravity or by hand, then forward checking with position bits can be used. Forward checking with position bits can detect an out-of-place mechanism, alert the operator, and prevent crashes. It is an excellent debug tool, for it allows the operator to use the manual-mode layer to help free product jams and so forth. Manual-mode features may not be needed for simple mechanisms but may be required when the mechanisms weigh more than you do.

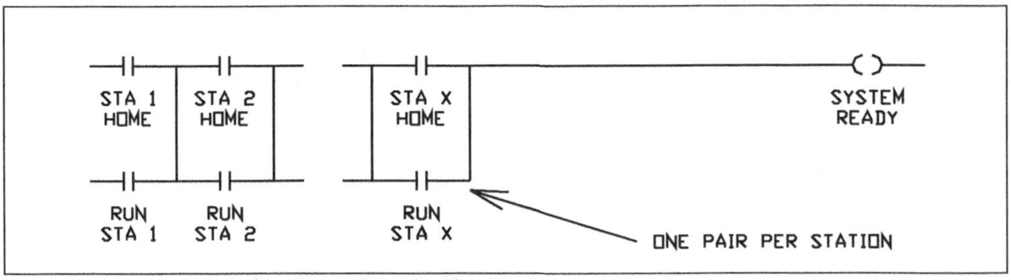

Figure 6-9. Typical System-Ready Rung

6.2.5 Station Reset

The *Station Reset* rung turns on when the sequence finishes. It is used to turn off all the sequence latches to ready the sequence for another cycle. A typical *Station Reset* rung is shown in Figure 6-10. *Station Reset* is triggered when the last task of the sequence is done, typically by the *Motion-Complete* bit from the last motion object. Notice that once the sequence is reset, the *Sequence Reset* bit turns off. As a result, the *Station Reset* bit is a pulse that is only on for one scan. Also notice that during the scan that the sequence is resetting, the sequence cannot operate.

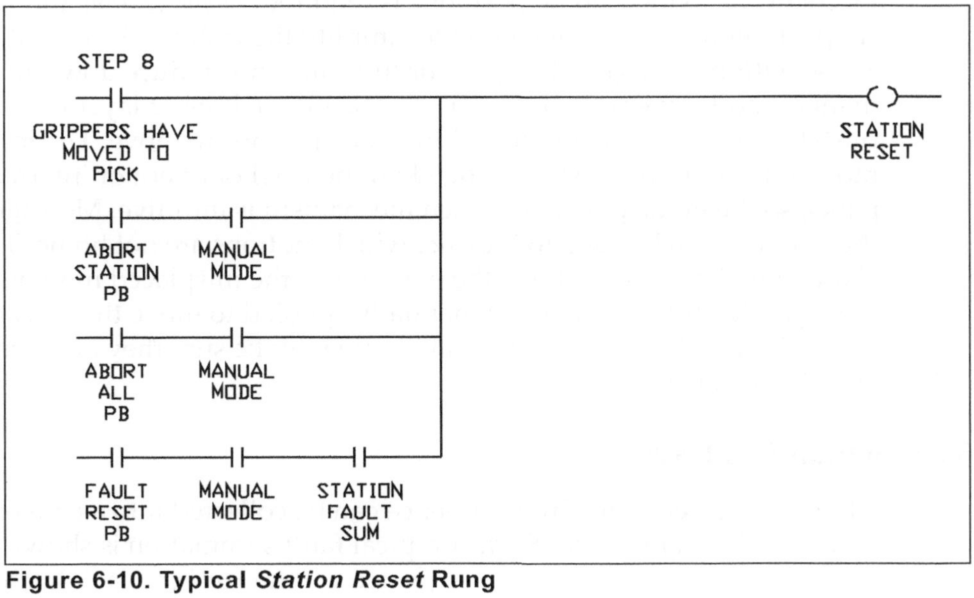

Figure 6-10. Typical *Station Reset* Rung

Alternate means of resetting a sequence should be provided. These other methods are provided in case things go wrong and to give the operator recovery options. Failed mechanisms and damaged sensors can prevent a sequence from finishing, so a means of aborting the sequence is needed. In some instances it might be faster to remove the damaged product, abort the sequence, and start over. Whatever the reason, a back-door method of resetting a sequence provides the guarantee that a sequence will always finish, by hook or by crook.

Several alternative reset methods are shown in Figure 6-10. The *Abort Station* push button aborts only the desired sequence. The *Abort All* push button resets all the sequences so the whole machine can start over. Both of these push buttons require the machine to be in manual mode to prevent aborting a running sequence and to make it harder to use in case of accidental initiation. The last branch in Figure 6-10 shows a method using the *Fault Reset* push button. This method will only abort a faulted sequence. Having unique push buttons to abort individual sequences is the best approach. The operator will always know the consequences of pressing the button and allows only the selected sequence to be aborted. The remaining sequences will wait until the faulted sequence is ready, then continue when the machine is restarted. Adding extra push buttons is not a problem when programmable operator interface devices (HMI) are used.

The mechanisms of an aborted sequence will most likely be in positions other than their home positions. At least one method of returning the mechanisms to their home positions is required. *Home Station* and *Home All* push buttons can be included to simplify the task for the operator. These buttons will save the operator time and thus reduce downtime. The manual-mode layer can also be used. The manual-mode layer includes push buttons to command individual mechanisms to move. The manual-mode push buttons can be arranged on the HMI or operator interface panel, so the home position of each mechanism is intuitive. Messages to the operator can be used to indicate which motions are not home. These "Not Home" messages direct the operator to the misplaced motions. Finally, solenoid valves can be manually spooled to move the mechanisms. If you plan on using the valves, then make sure they are accessible and clearly labeled.

6.2.6 Station Fault Sum

All the faults generated by a sequence can be collected into a convenient bit called the *Station Fault Sum*. A typical fault summation is shown in Figure 6-11. It is used to freeze the sequence via the *Station Enable* latch. It is also added with all the other fault summations to create a *System Fault Sum*. The system control (*Run System*) is turned off and locked out when the *System Fault Sum* is on. Without the system control, all the other sequences are prevented from restarting. Fault summations are useful when debugging the equipment because they list all the faults. You can find out which fault is on by starting with the *System Fault Sum*, then working backward to the individual fault. A better method is to get the message logic working early in the machine startup process. The message logic will tell you which fault is on. Message logic is discussed in Chapter 13.

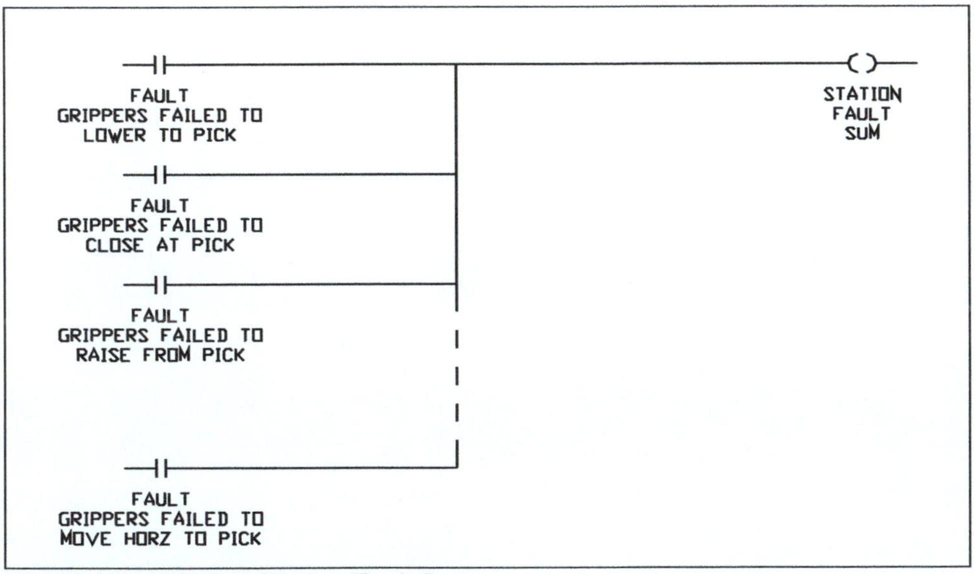

Figure 6-11. Typical *Station Fault* Sum

6.3 Conclusion

The Sequence Control rungs encapsulate the sequence and makes it self-contained. They also provide information to the system control and other sequences. Self-contained sequences can continue to finish their processes when another sequence develops trouble and stops. This reduces down-time by containing the problem. Self-containment also allows each sequence to operate by itself to save us time when debugging and commissioning the equipment. Stand-alone operation will be shown in chapter 7, where logic is added to allow stepping and single cycling of sequences.

7 Step and Single Cycle

7.0 Introduction

Chapter 6, Controlling Sequences, discusses how two latches control the behavior of a sequence. The *Run Station* latch is used to start a sequence and to remember that the sequence was started. The *Station Enable* latch is used to allow a sequence to execute and cause motion or to freeze it and prevent motion. This chapter will modify these two latches to allow stand-alone execution of a sequence so we can operate just a portion of the machine by itself. The modification will allow a sequence to execute one step at a time (stepping) or to allow it to execute completely (single cycle). A method of mode selection will also be discussed, since we now have to contend with step and single-cycle modes in addition to manual and automatic.

An important phase in the life of an automated machine is startup/commissioning. Startup begins immediately after the machine is built, and the controlling logic is downloaded into it for the first time. This phase is usually called *debug*, but I prefer calling it *startup* or *commissioning* because "debug" implies original sin. I like to think I was thorough enough in programming that after adjusting a few sensors and timers, the machine is good to go; at least this is the goal. Experience, on the other hand, tells me to build in some tools to assist the startup effort. As it turns out, these same tools are also useful to the end user in operating and maintaining the equipment.

7.1 Why Have Stepping?

One of the startup tools is the ability to step through a sequence. Stepping allows execution of a sequence one step at a time on command from the user. Manual mode is used to operate each mechanism to adjust flow controls, hard stops, and sensors. Manual mode is also used to walk product through the machine to check alignment, gripper pressures, and so forth. However, operating individual mechanisms does not check the sequence; stepping does. It is a good bridge to test a section of a machine before running it at full speed (single cycle), then in full automatic mode with all the other sections.

7.2 Why Have Single Cycle?

Another tool is the ability to single cycle a sequence. Single cycling a sequence operates all the associated mechanisms the same way the machine would while running in automatic mode. In fact, they both used the same sequence. The difference between single cycling and automatic operations is the ability to select a small section of the machine to test. Single cycling is used after checking the mechanisms in manual mode, then the sequence in step mode. Single cycling tests mechanism and product interactions at operating speeds because they act differently when momentum is involved. It is also used to check cycle time.

Manual mode, step mode, and single cycle allow startup of a machine in an orderly manner, one machine section at a time. It adds up as time saved starting up and commissioning the equipment. As a result, it saves the machine builder money, ships the machine sooner, and makes you—the programmer—look good. The end user can start realizing the capital investment in the equipment sooner; thus he or she saves money and is more likely to use you again.

7.3 Mode Selection

Separating the possible machine actions into operating modes makes it clear to the users how the machine will respond to requests. When the machine is running in automatic mode, we do not want the operator commanding individual motions, so the operator is locked out. When the machine is in manual mode, the operator knows that only the chosen mechanism will move, so sequence execution is locked out. Two additional modes are required to allow stepping and single cycling.

The addition of stepping and single cycle makes mode selection more complicated. It is now more than one bit being on or off to select between manual and automatic mode. The following are the three requirements of mode control:

1. The machine can be in only one mode at a time.

2. There should be a clear indication of which mode the machine is in.

3. Changing modes should not cause problems.

Mode selection and display can be accomplished with a multiple-state HMI push button. This button will send a signal to the machine controller whenever it is pressed. The machine controller then tells the push button to indicate the selected mode. Figure 7-1 shows a binary counter circuit that uses *Mode 1* and *Mode 2* latches to count between 0 and 3. The circuit counts by one each time the mode selection push button is pressed. The *Mode 1* and *Mode 2* latches can be deciphered into the possible machine modes as tabulated in Figure 7-2. The decoding logic is shown in Figure 7-3. The individual mode bits—manual, step, single, and auto—are used by the machine logic and fed back to the mode selector push button to display the correct words.

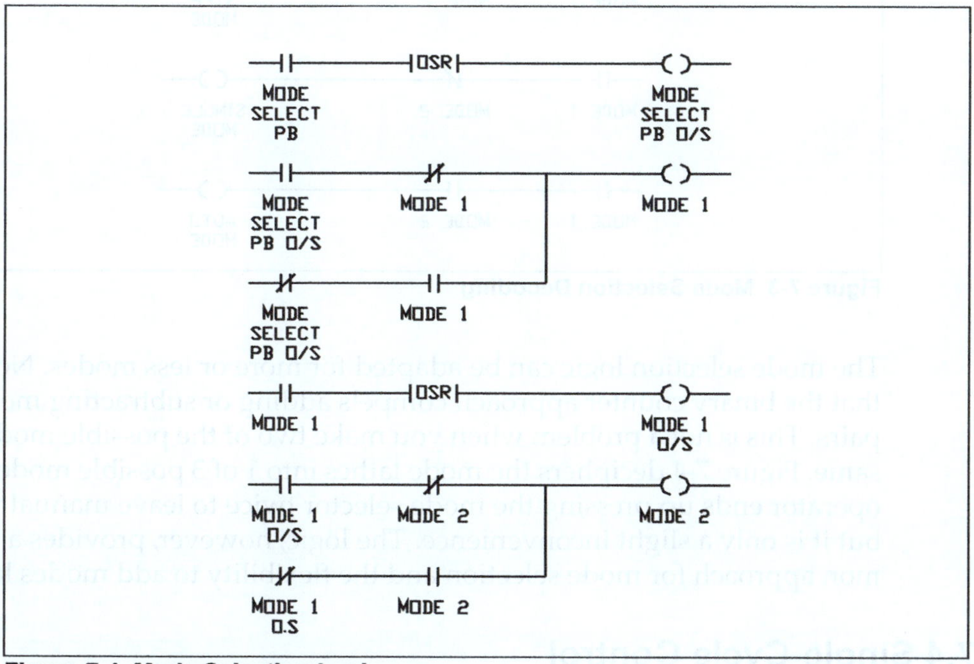

Figure 7-1. Mode Selection Logic

Both the *Mode 1* and *Mode 2* latches will be off when the machine controller initializes. This will make manual mode the default mode. Manual makes a useful default mode, since some type of manual action is usually required to ready the machine. Some processors may not reset their internal coils, so other logic, such as a *First Scan* bit, will be required to select a default mode.

Mode 1	Mode 2	Machine Mode
Off	Off	Manual Mode
Off	On	Step Mode
On	Off	Single Cycle Mode
On	On	Automatic Mode

Figure 7-2. Mode Selection Truth Table

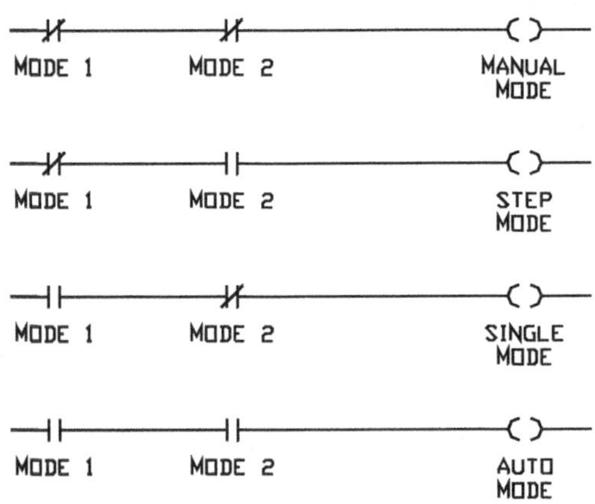

Figure 7-3. Mode Selection Decoding

The mode selection logic can be adapted for more or less modes. Notice that the binary counter approach compels adding or subtracting modes in pairs. This is not a problem when you make two of the possible modes the same. Figure 7-4 deciphers the mode lathes into 1 of 3 possible modes. The operator ends up pressing the mode selector twice to leave manual mode, but it is only a slight inconvenience. The logic, however, provides a common approach for mode selection and the flexibility to add modes later.

7.4 Single-Cycle Control

Single cycling is allowing a sequence to execute from start to finish just once each time the user desires. This is already inherent in the sequence control. A typical *Run Station* latch is shown in Figure 7-5. If you want a sequence to run continuously over and over again, just keep the start trigger on to set the *Run Station* latch after *Station Reset* turns it off. If you only want a single execution of a sequence, then remove the start trigger after *Run Station* turns on to prevent the sequence from restarting. Only two things are needed to complete the single-cycle logic: The first is an opera-

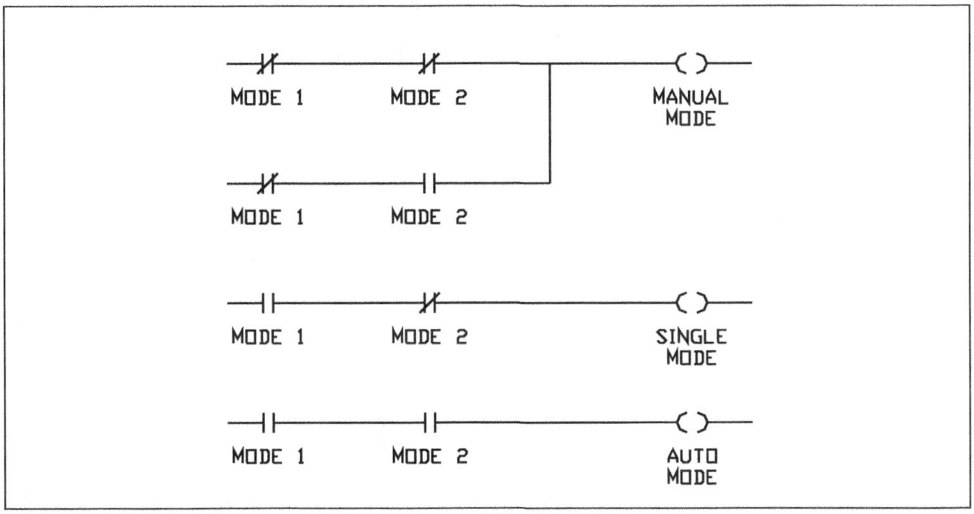

Figure 7-4. Creating an Odd Number of Modes

tor input to initiate the action, and the second is interlocking to make sure it is safe to start.

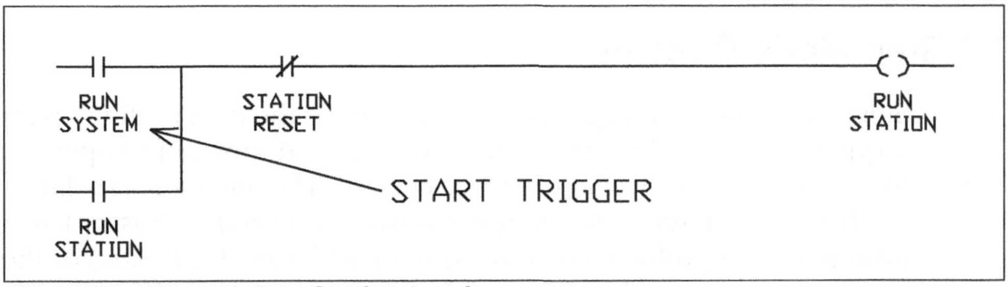

Figure 7-5. Typical *Run Station* Latch

On a simple machine, using one sequence the *Cycle Start* push button conditioned with the appropriate operating mode bit can be used to initiate the execution of a single step, to initiate a single cycle, or to start a full automatic cycle. However, it is always better to avoid any confusion by giving each machine action its own push button. On multiple-station machines, separate push buttons to single cycle and step each station should be provided. An HMI is typically used on larger machines because they are easy to use, easy to program, and can hold many operator interface objects.

Before a machine can be started in automatic mode, it needs to be in its home position. This guarantees the mechanisms are at a known starting point and synchronized with the sequence. Similarly, a station needs to be at a known starting point before it can start a single cycle, but it also has to be concerned about its neighbors. Additional interlocks beside the station home summation are required if a neighboring station can interfere with the single cycle. Typically, when working on one station we do not need to concern ourselves with the other stations, and they do not necessarily have to be home themselves.

A protruding mechanism from a neighbor could collide with the station, and it should be retracted before allowing the station to start. This can be accomplished by using a sensor that indicates the mechanism is retracted with the *Single-Cycle* push button. The protruding mechanism may be clear when the station starts, but it could be commanded to move later. To make sure this does not happen, the neighboring sequence can be locked out when the station is running. The station can also be prevented from single cycling when neighboring sequences are running.

To single cycle a pick/place station on a dial machine, make sure the station is in its home position, the dial sequence is not running, the dial is in dwell, and there is no product in the dial's nest. The pick/place unit can now be commanded to place a part on the dial over and over again to test its operation. You will have to remove the product from the dial's nest each time, but the machine is protected if you do not.

7.5 Step Mode Control

Stepping through a sequence involves letting the next motion object in the sequence operate. The trick is to let only one motion object operate. This too is also inherent in the sequence control. The *Station Enable* latch is normally turned on when the system control turns on (*Run System*). It stays on until there is a problem with the sequence (*Station Fault Sum*) or the machine is not capable of running or is not safe to do so (*System Enable*). A typical *Station Enable* latch is shown in Figure 7-6.

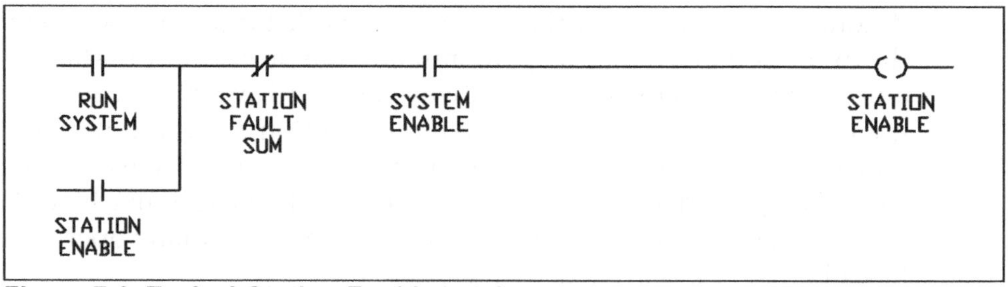

Figure 7-6. Typical *Station Enable* Latch

Suppose the *Station Enable* latch turns on for only one scan. The last finished motion object will then trigger the next motion object to start. The inputs to signal the completion of the started motion object will not be scanned by the processor until the next scan, at which time the *Station*

Enable latch will be off. As a result, only one motion object will execute. Incorporating stepping is just a matter of turning the *Station Enable* latch on for one scan each time an operator presses the *Step* push button. There will also be interlocking, as discussed for single cycling.

7.6 Modified Sequence Control

Figures 7-7 through 7-9 show the modified sequence control rungs needed to incorporate stepping and single cycling. The rungs are shown to illustrate their placement in the machine control program.

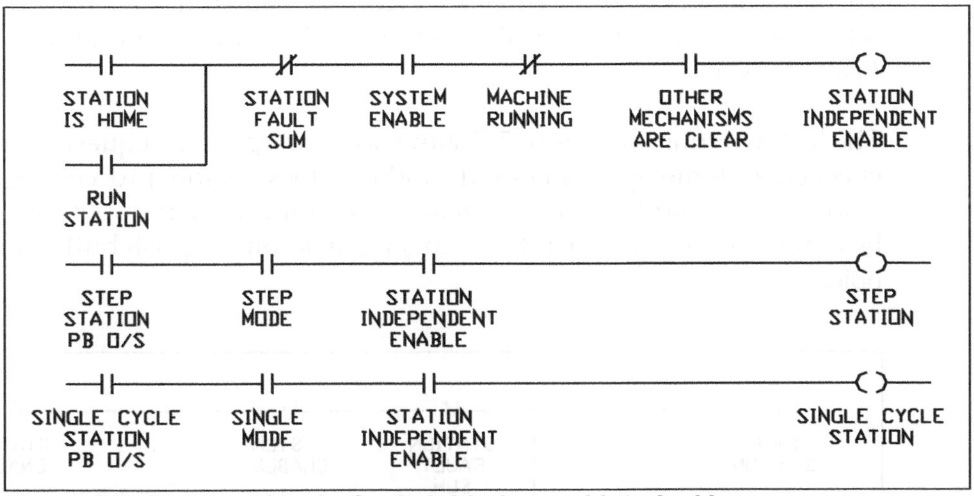

Figure 7-7. Step and Single-Cycle Initiation and Interlocking

The *Station Independent Enable* rung sums up all the conditions and interlocks required to allow stepping or a single cycle. A separate enable rung for each station is used, since these conditions are unique for each station and different from those required to start the machine as a whole in automatic mode.

The first interlock, *Station Is Home*, checks to make sure the station mechanisms are synchronized with the sequence. The mechanisms must be at their starting point when the sequence is first started. The *Run Station* bit bypasses the *Station Is Home* bit if the sequence is already started, as it would be if the sequence is frozen due to a fault or is between steps. It is assumed that the mechanisms and sequence remain synchronized after the sequence is started. Forward checking and position bits are used when there is a good possibility mechanisms will move when a sequence is frozen.

The next two interlocks, *Station Fault Sum* and *System Enable,* check to make sure the station has no problems and that it is safe to operate and capable of operating.

The *Machine Running* interlock makes sure no other sequence is executing and then locks them out when the sequence starts. This will prevent mechanisms that are currently clear from later moving and interfering. Only allowing one sequence to execute at a time makes the machine safer, in case two groups of people are working on the equipment. How to construct a *Machine Running* summation is discussed below.

The last interlock, *Other Mechanisms Are Clear,* illustrates where all other interlocks are placed. Examples include *Dial In Dwell* and *Dial Nest Is Empty* sensors.

The last two rungs in Figure 7-7 show how the operator request signals are checked with the operating mode and interlocks before proceeding to the *Station Enable* and *Run Station* latches. The inputs from the operator should be a pulse in case the operator's finger is held on the push button or HMI object.

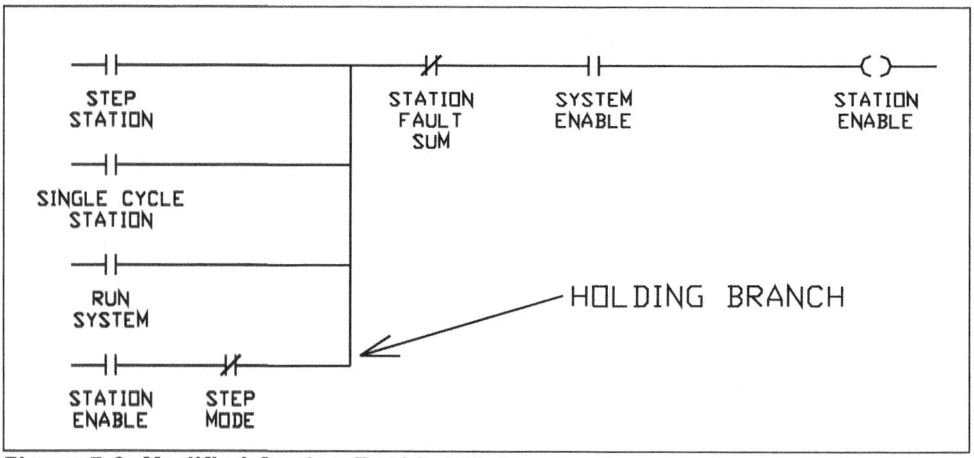

Figure 7-8. Modified *Station Enable* Latch

The *Run System, Single Cycle Station,* and *Step Station* bits are used to turn on both the *Station Enable* and *Run Station* latches, depending on which mode the machine is in. The *Step Mode* bit is used to keep the holding branch of the *Station Enable* latch from working. As a result, the *Station Enable* latch will pulse on for one scan when the *Step Station* bit is pulsed on.

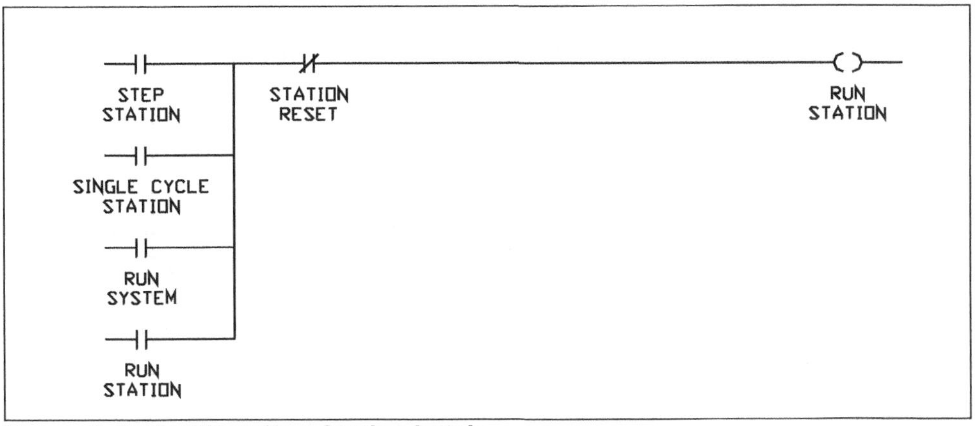

Figure 7-9. Modified *Run Station* Latch

7.7 Mix and Match Modes

Notice that a sequence is frozen in place if it faults out, regardless of which mode the machine is in. The sequence is also frozen in place between steps, in step mode. Also notice that a sequence can be restarted in any mode, and it does not need to be the mode it stopped in. This means a faulted sequence, or a sequence stepped partially through, can be restarted with the rest of the machine (automatic mode), by itself (single-cycle mode), or by itself one step at a time (step mode). Many times it is useful to walk the mechanisms of a station through a critical operation in step mode and then have the machine finish the sequence in single-cycle mode.

Although machine operations were not intended to be mixed and matched between modes, this does provide a more flexible machine. As a machine builder, I find the versatility very useful, especially when working on new processes and tooling. I have also seen operators using the choices to restart the equipment as appropriate to the current situation, thus reducing downtime.

7.7.1 Bump-Less Transfer

Avoiding problems caused by changing modes is sometimes referred to as bump-less transfer, a term originally used by PID controllers to describe how the percent output of the controller remains the same when the operating mode changes. This basic principle was applied to the output logic to switch between automatic and manual modes without the outputs changing states. Even if a mechanism does not present a hazard to the operator, its movement on a mode change does not look good. It will have the effect of lowering everybody's confidence in the equipment. In general, mechanisms should move only when specifically told to do so by the operator. This is achieved by pressing buttons with labels such as *Cycle*

Start, *Home Station 5*, or *Extend Press Clamps* not from side effects. This will present a predictable and thus safer machine to the end users.

Another possible problem occurs when the operator selects manual mode while the machine is running automatically. Since manual mode allows an operator to command individual mechanisms, it is conceivable that the machine and the operator can end up commanding motions at the same time. Can you say "crash"? It is not enough to simply group and select user requests into modes, but it is necessary to check if it is actually safe to allow an action. This is what the *Station Independent Enable* rung shown in Figure 7-7 does.

The machine can be cycle stopped on a mode change and the manual mode functions can be locked out until the machine stops, to prevent the operator and machine from simultaneously commanding mechanisms. The *Mode Select PB O/S* bit, shown in Figure 7-1, can be used to cycle stop the machine by using it to unlatch the system control (*Run System*). The *Manual Enable* rung, as discussed in Chapter 3 on output logic and shown in Figure 7-10, is modified to include the *Machine Running* bit. This will prevent all manual actions if any sequence is executing.

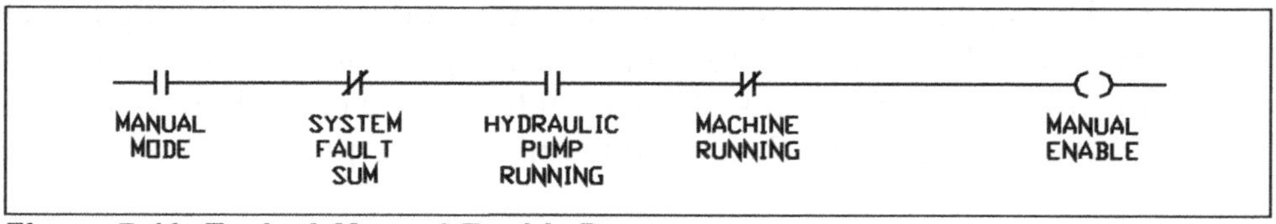

Figure 7-10. Typical *Manual Enable* Rung

7.8 Machine Running

The *Machine Running* bit is a very useful bit and is essential to the safe operation of the machine. It is used as an interlock, to operate pilot lights, and to lock/unlock guard doors. It is basically the summation of all the individual *Station Running* bits. A typical *Station Running* bit is shown in Figure 7-11.

Figure 7-11. Typical *Station Running* Rung

Station Running is on whenever the associated sequence is executing in either automatic or single-cycle mode. The problem is, *Station Running* is only on for one logic scan, when a sequence executes in step mode. The respective motion may take several seconds to complete and will render *Machine Running* a poor interlock in step mode. The solution is to use a delay off timer to stretch the step mode pulse long enough for the slowest motion to finish. The delay off timer is incorporated into the *Machine Running* summation instead of the individual *Station Running* rungs, as shown in Figure 7-12.

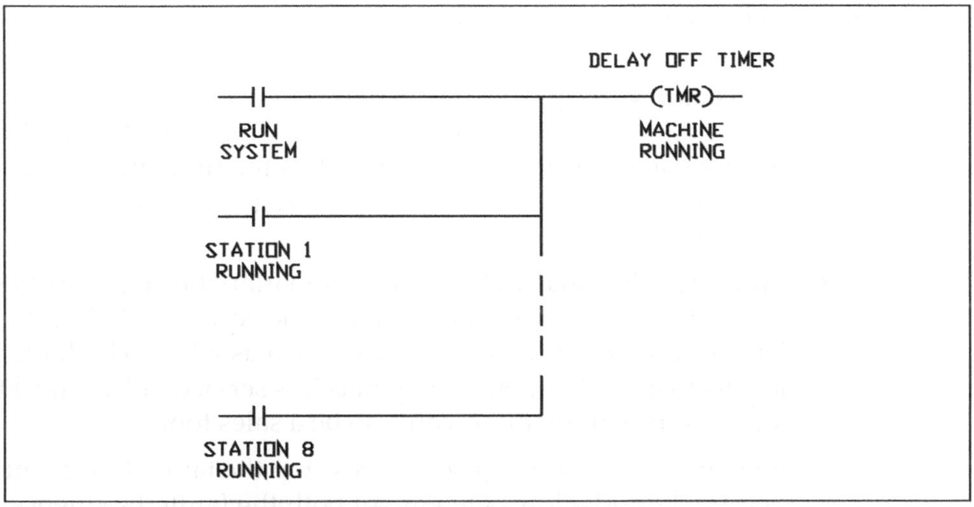

Figure 7-12. Typical *Machine Running* Summation

The system control, *Run System,* is also added to the *Machine Running* summation to turn it on immediately after the operator starts the machine. Just in case the sequences do not start immediately, at least the *In Cycle* pilot light is on and the guards are locked. This situation may arise where the machine is started empty, and it has to wait for product.

7.9 In Cycle

The *In Cycle* pilot light illuminates when the machine is running. It flashes to indicate intermittent machine operation, such as when the machine is cycle stopping, stepping, or single cycling. The light blinks whenever a sequence is executing, as indicated by the *Machine Running* summation, and when the *Run System* bit is off. The pilot light circuit is shown in Figure 7-13.

7.10 Fault Recovery, Debug, and Startup Tools

As a review, let's look at the tools the operator and the programmer have to diagnose and recover from problems. These tools share the goal of reducing downtime and saving product.

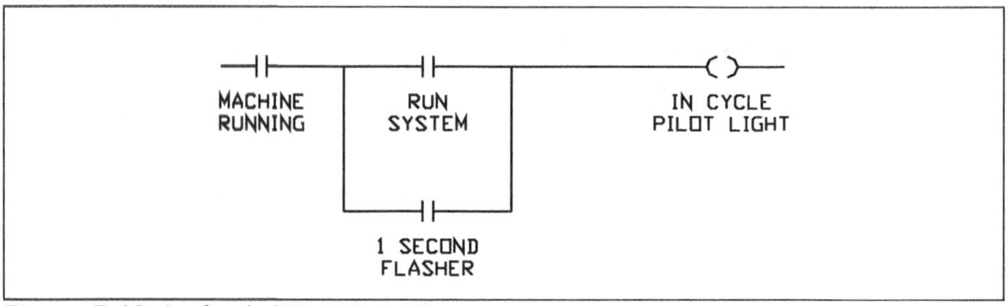

Figure 7-13. *In Cycle* Pilot Light Rung

- **Faults and Messages.** Faults stop machine operations to prevent further damage to the equipment and/or product. Faults also provide messages to alert the operator. It is the first line of defense. Fault messages are also useful to the programmer because they can be used during machine startup.
- **Equipment Manuals.** The more information the operators have, the better they can operate and maintain the equipment. Additional detail of each fault can be included, such as what to look for and how to recover. The benefit to you is less service calls. A tradition of good equipment manuals can also be a sales tool.
- **Freezing Logic.** Freezing sequences stop the faulted mechanisms in their tracks and allows viewing of both the faulted sequence logic and the faulted mechanisms. The determination of what went wrong is easier when you can see the conditions causing the trouble.
- **Cycle Stop Remaining Stations.** When one sequence stops due to a fault, the remaining sequences will cycle stop to a more stable state. This allows entry into the machine without the loss of motive power (electricity and air) causing problems with the non-faulted stations.
- **Resuming.** Frozen logic has the ability to resume. If the problem can be fixed, the fault can be reset and the machine restarted. These are recoverable faults, such as minor proximity switch adjustments or reseating misbehaving product.
- **Aborting.** If the problem is not easy to fix, the sequence can be aborted. Busted proximity switches and worn-out cylinders are examples of nonrecoverable faults. Aborting a sequence might even be faster for the operator. Product that is out of specification can jam inside the machine. Removing the product and starting the station over can be very fast. If the station is placing an inexpensive component into a larger assembly then yank, reset, and go might be the preferred recovery method. Aborting also means homing, so make homing easy for the operator.

- **Manual Mode.** Manual mode can command mechanisms to complete the process and save the product. The logic used to determine the pass/fail status of the product should not be dependent on sequence execution to allow this. Manual mode is also used to place the mechanisms in their home positions. The programmer and maintenance people use manual mode for setup and debugging mechanisms. Flow controls are set by commanding a cylinder to move back and forth. Alignment is checked by walking the product through the machine in manual mode.
- **Forward Checking.** Forward checking with position bits allows resumption of automatic operations after moving mechanisms in manual mode or by hand. If the mechanisms or product are too hard to move by hand, then the operator will need to use the manual mode layer. A person may not remember to return a mechanism back where it belongs but at least the machine is protected and the operator is told which mechanism to re-position.
- **HMI Layout.** How the operator interface devices are presented to the operator has a big impact on effectiveness, especially to new operators. The operator interface devices can range from discrete push buttons and analog meters to HMI units. The selections of color, location, size, and text should all be considered.
- **Independent Station Operations.** With each machine section under control by its own sequence, a divide-and-conquer approach is possible. This allows concentration on small pieces of the machine to speed logic development, setup, debug, and operation.
- **Stepping.** Stepping is useful during equipment setup and debug. The sequence can be walked through to check the sequence, each mechanism for alignment, sensor placement, product interaction, and so forth.
- **Single Cycling.** Single cycling lets a sequence operate just once to check the dynamics of the mechanisms and product interaction at operational speeds. Mechanism and product act differently when momentum is involved.

7.10.1 General Recover Theory

Situations will arise that will require special handling because nothing can be completely thought out and planned for. The users of the equipment can always find a way of proceeding when setup, testing, and recovery tools are provided. Operators have eyes, ears, and brains more powerful than any combination of sensors or processors that we can build into a machine. Use the operator to complete the control loop in these special cases by providing information and options to the operator.

The general recovery theory goes as follows:

1. Protect the operator at all times. A few methods to protect the operator include training, guards, safety circuits, lock-out/tag-out points, warning signs, and equipment manuals.
2. Provide the operator with useful information, such as having the machine alert the operator to the problem and descriptive equipment manuals. The equipment manual can describe the individual fault messages in more detail and outline recovery options.
3. Bring the operator into the machine control loop. Let the operator inspect the problem and determine what to do. The best inspection sensor available is a well-trained operator, so make use of this valuable resource. The first step in any recovery procedure is visual inspection; always stress this point. Here again the equipment manual can help by listing items to look for.
4. Provide the operator methods of recovery. No matter how screwed up the machine gets, always provide a way for the operator to recover through the operator interface devices. Plugging in a support computer is not an acceptable method and will reflect poorly on the programmer.
5. Protect the machine from inadvertent operator actions. Protecting the machine in this case is referred to as *interlocking,* as was discussed in Chapter 3 on output logic. It is easy to press the wrong button, especially when an HMI is used. The worst that should happen is a few more recovery steps, not machine damage.

8 System Control

8.0 Introduction

At this point in the development of the machine-control logic, we have complete sequences, with each sequence operating a group of mechanisms on the equipment. The mechanism groups will be referred to as *stations* to reflect how these mechanisms are actually located on the equipment, and how the mechanisms within a group work together to perform an individual process on the equipment. Each sequence can be walked through step by step in step mode, executed just once on each start command in single-cycle mode, executed just once on each start command from another sequence in automatic mode, or operated continuously by keeping the start command on in automatic mode.

The last two chapters on controlling sequences talked briefly about the system control and how it started and enabled the sequences. This chapter will discuss the system control in detail and explain the hierarchy of the control logic. Before discussing the system-control functions, let's review how sequences are controlled.

8.1 Summary of Sequence Control

Each sequence, no matter if it is independent, a parent, or a child, is controlled with three basic rungs. These are the *Station Independent Enable*, *Station Enable*, and the *Run Station* rungs. Figure 8-1 provides a quick summary of how the *Station Enable* and *Run Station* latches are interpreted to determine the status of the sequence. In a nutshell, *Enable Station* lets a sequence execute or not and *Run Station* starts the execution and remembers that it was started.

Run Station	Station Enable	State of Sequence
Off	Off	Sequence Off and Not Ready
Off	On	Sequence Off and Ready
On	Off	Sequence Frozen
On	On	Sequence Runing

Figure 8-1. *Station Enable* and *Run Station* Latches

The sequences are autonomous, in that each one can operated independently, provided the appropriate interlocks are valid. The outputs are even autonomous with the appropriate interlocks in manual mode. To finish the review of sequence control, let's look at what starts the sequences and the interlocks that are required to be satisfied before the sequence can execute for each operating mode.

8.1.1 Manual Mode

Function

- Operate each motion independently for equipment setup and debug.

Command To Start Sequence

- None.

Command To Enable Sequence

- None.

Interlocks

- Outputs are safe and capable of operating—*Manual Enable* from the system control.
- System Interlocks, which the *Manual Enable* command requires.
- Appropriate sensors to prevent crashes.

8.1.2 Step Mode

Function

- Walk sequence through one step at a time to check sequence.

Command To Start Sequence

- Step push button.

Command To Enable Sequence

- Step push button.

Interlocks

- No faults exist—*Station Fault Sum* from the sequence.
- Sequence is home or previously started—*Sequence Home Sum* from the sequence and *Run Station* from the sequence.
- Sequence is safe and capable of execution—*System Enable* from the system control.
- Mechanisms from neighboring stations are clear—appropriate sensors to prevent crashes.
- No other station sequences are executing—*Machine Running* from the system control. This makes the machine safer by allowing only one sequence to operate at a time, in case more than one person is working on the equipment, and prevents neighboring sequences from sending mechanisms into the station that is running.
- System Interlocks, which the *System Enable* command requires.

8.1.3 Single-Cycle Mode

Function

- Operate sequence at rated speed for setup and debug.

Command To Start Sequence

- Single-cycle push button.

Command To Enable Sequence

- Single-cycle push button.

Interlocks

- No faults exist—*Station Fault Sum* from the sequence.
- Sequence is home or previously started—*Sequence Home Sum* from the sequence and *Run Station* from the sequence.
- Sequence is safe and capable of execution—*System Enable* from the system control.

- Mechanisms from neighboring stations are clear—appropriate sensors to prevent crashes.
- No other station sequences are executing—*Machine Running* from the system control. This makes the machine safer by allowing only one sequence to operate at a time, in case more than one person is working on the equipment, and prevents neighboring sequences from sending mechanisms into the station that is running.
- System Interlocks, which the *System Enable* command requires.

8.1.4 Automatic Mode

Function

- Operate equipment for production.

Command To Start Sequence

- *Run System* from the system control for independent or parent sequences, or
- Start trigger from a parent sequence for child sequences.

Command To Enable Sequence

- *Run System* from the system control.

Interlocks

- No faults exist—*Station Fault Sum* from the sequence.
- Sequence is safe and capable of execution—*Manual Enable* from the system control.
- System Interlocks, which the *Run System* and *System Enable* commands require.

The *Manual Enable*, *Run System* and the *System Enable* are start commands and interlocks from the system control. We need to discuss the relationship between the sequences and the system control before we can discuss how the system control signals are controlled and interlocked.

8.2 Control Hierarchy

The machine control is divided into layers to allow the machine motions to operate individually (manual mode), in subgroups (stations and sequences), and all together (the whole machine and the system control). Independent operation allows faster setup, debugging, and testing of the equipment on a motion-by-motion and station-by-station basis. The machine control is also layered or divided to provide encapsulation. Encapsulation allows each sequence and layer to be programmed independently with a minimal need to see how the other sequences are programmed. This speeds program development and makes the code more consistent. The relationship between the sequences and the system control forms a hierarchy with a definite pecking order of who controls whom. The machine-control hierarchy has the following four layers:

8.2.1 System Control Layer

The system control layer has the highest priority. It starts and stops the machine as a single entity and controls mode selection. It monitors the safety systems to see if it is safe to operate the motions. It monitors the motive power to let motions and sequences operate only when they are capable of working. It includes the control of the devices common to the entire machine, such as hydraulic pumps, vacuum pumps, dust collectors, overhead stacklights, and door locks. I refer to these common devices as *auxiliary equipment*, which either needs to be running before the equipment is started or started by the system control.

8.2.2 Parent and Independent Control Layer

Parent and independent sequences have the next-highest priority. These sequences are started when the system control, *Run System*, starts and continue to operate as long as the system control stays on or a fault within the sequence is generated. Any fault also turns the system control, *Run System*, off. A small machine with a single process only has one independent sequence. A larger machine may have several independent sequences and a parent sequence. The sequence for the dial on a dial machine is a parent sequence.

8.2.3 Child Control Layer

Child sequences are third in the hierarchy. These sequences start when a parent sequence triggers and execute until finished or until a fault with the sequence is generated. The sequences controlling the stations on a dial machine are child sequences.

8.2.4 Manual Mode Layer

The priority of the manual mode layer is on par with the independent, parent, and child layers, in that it depends on the system layer to function. The manual mode layer is distinct from the sequence layers because the outputs can receive commands only from the operator or from the sequences and never from both at the same time. Having separate operating modes enforces the distinction. The manual mode layer allows individual motions to be commanded by the operator with the minimal quantity of interlocks. The sequences do not even need to be programmed to allow manual mode operations.

8.2.5 Miscellaneous Logic

The logic in this category does not necessarily have any priority in the control hierarchy, but a separate category is needed to indicate there is more logic to machine control than just sequences and a system layer. Controlling operator messages and compiling production statistics are examples of the functionality of the logic included in this category. The message logic is discussed in Chapter 13. You are on your own for production statistics, but you can use signals from the system layer and the sequences to control timers and increment counters.

8.3 System Control

The system control handles the interface with the operator, to provide orderly starting and stopping of the equipment. It monitors the safety systems to allow motion only when it is safe to do so. It also monitors the motive power to allow the initiation of motion only when the motions are capable of working. The system control does this with three main rungs of logic: *System Enable*, *Run System*, and *Manual Enable*. Other rungs of logic such as *System Ready* and *System Fault Sum* support the main rungs. Let's look at what makes up each of these rungs.

8.3.1 System Enable

Figure 8-2 shows a typical *System Enable* rung used to monitor the equipment's safety systems and motive power and includes inputs from the master circuit and a pneumatic pressure switch. The safety circuits take many forms but will be referred to collectively as the *master circuit*. The prime movers are provided with whatever motive power they need (electrical and pneumatic) when the master circuit is energized. The master circuit is de-energized to remove all motive power when the machine becomes unsafe for the operator, such as when a guard door is open and exposes mechanisms that could move. The master circuit tells the control logic if the machine is safe or not and when motive power is available. The pneumatic supply is monitored with a pressure switch because it can be lost when the dump valve is operated by hand or during heavy usage when the pressure drops in addition to when the master is de-energized.

Figures 8-3 and 8-4 show how to construct the logic for the *Master On Plus Time* and the *Air Is On Plus Time* contacts in the *System Enable* rung. Time delays are used to allow pneumatic systems to pressurize, servo controllers to initialize, and so forth. The time delays begin after the respective faults are reset to allow the operator to view and acknowledge the condi-

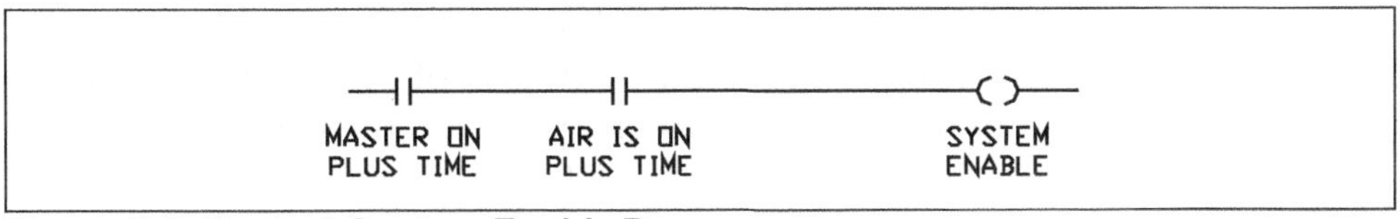

Figure 8-2. Typical *System Enable* Rung

tion. Remember that faults can only be reset when the faulting conditions are gone.

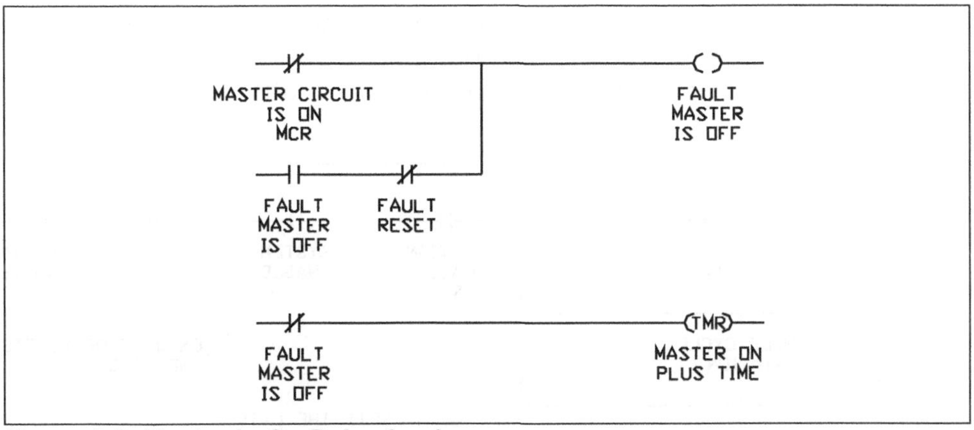

Figure 8-3. Master Is On Delay Logic

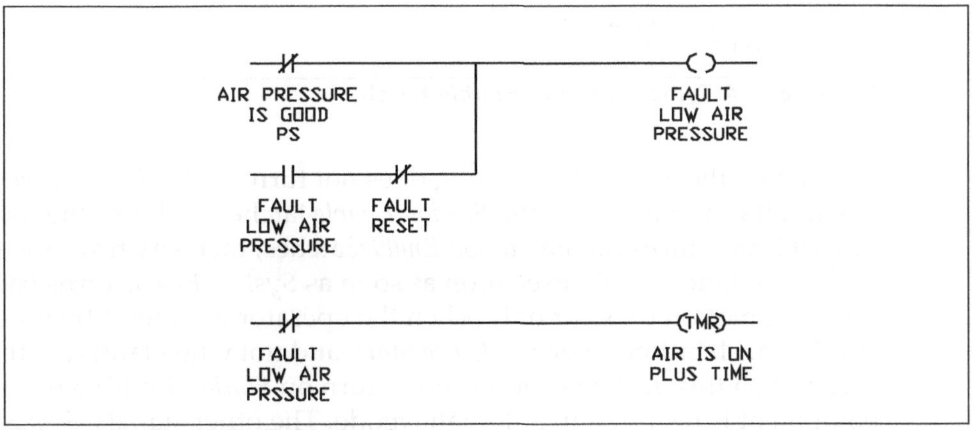

Figure 8-4. Air Is On Delay Logic

A hydraulic pump is a common resource that is included in the *System Enable* rung if the operator is required to start it before starting the machine. The contact inserted into the *System Enable* rung indicates the pump is running and there are no problems with it. The items making up the contact include input from the motor starter and faults monitoring the overload relay, fluid temperature, fluid level, and status of the pressure drop across the filter (clogged filter). In a similar manner, other auxiliary equipment that is essential to the operation of the equipment is monitored and included in the *System Enable* rung, when the operator is required to start these subsystems. Examples include vacuum pumps and dust collectors.

The *System Enable* allows motion by controlling the execution of all the sequences though the *Station Enable* latch of each sequence. A typical *Station Enable* latch is shown in Figure 8-5. Sequences are allowed to execute when the *System Enable* is on. The sequences are frozen (execution halted) when the *System Enable* is off. The prime movers controlled by the sequence cannot function without motive power, so it is best to halt execution to prevent the generation of sequence faults.

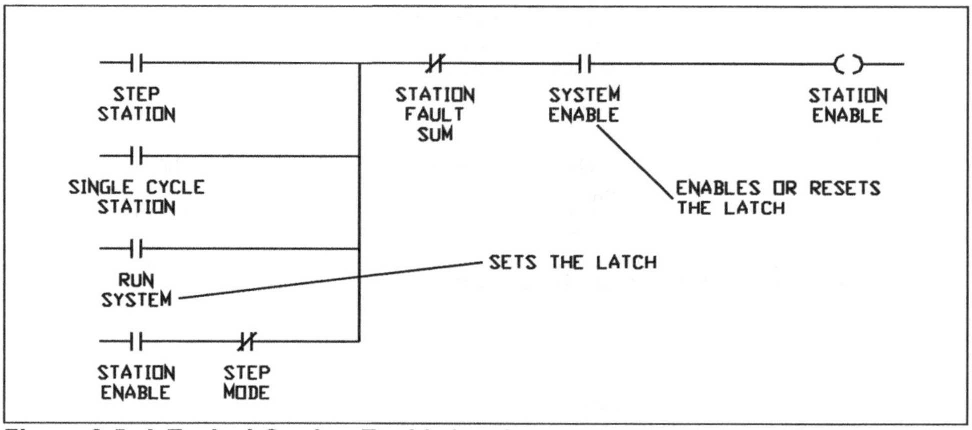

Figure 8-5. A Typical *Station Enable* Latch

Notice that the *System Enable* rung does not turn on the *Station Enable* latches. It can only enable the *Station Enable* latches and turn them off. If *System Enable* turns on the *Station Enable* latches, then any frozen sequence will immediately begin execution as soon as *System Enable* turns back on. We want motion to occur only when the operator initiates it by pressing specific push buttons, such as *Cycle Start*, and not when faults are reset. To accomplish this, *Run System* is used to turn on *Station Enable* when the equipment is restarted in automatic mode. The other signals, *Step Station* and *Single Cycle Station*, do this in the other operating modes.

8.3.2 Run System

The *Run System* latch is used to start and stop the sequences in automatic mode. A typical *Run System* latch is shown in Figure 8-6.

The *Run System* latch will turn on, providing the following conditions are met:

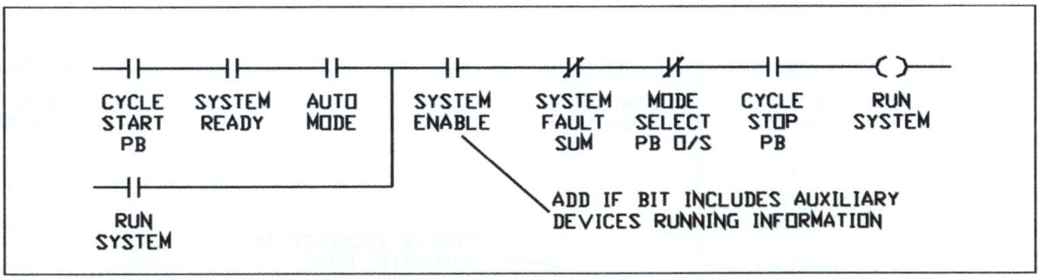

Figure 8-6. Typical *Run System* Latch

- The operator presses the *Cycle Start* push button.
- The machine is in Automatic Mode.
- The machine is capable of operating, and any auxiliary equipment the operator is required to start is operating—*System Enable.*
- There are no faults—*System Fault Summation.*
- All the stations are either home or ready to resume—*System Ready.*

The *Run System* latch stays on until any one of the following events occur:

- Operator presses the *Cycle Stop* push button.
- Any fault is generated—*System Fault Summation.*
- The operating mode changes—*Mode Select PB O/S*. This bit detects when the mode changes and its construction is shown in Figure 7.1 (Chapter 7).
- Loss of motive power—*System Enable*. Loss of motive power typically turns the *Run System* latch off through the *System Fault Summation* with the *Fault Master Is Off* and *Fault Low Air Pressure* faults.
- Completion of special modes, such as Purge or Clear Out, used to empty the machine of all component parts before it stops.

The *Run System* latch starts the machine by turning on the *Run Station* latches of all the independent and parent sequences. A typical *Run Station* latch for a parent sequence is shown in Figure 8-7. The child sequences are started by the parent sequences. A typical *Run Station* latch for a child sequence is shown in Figure 8-8. *Run System* also enables all the sequences by turning on their *Station Enable* latches as in Figure 8-5. Any frozen sequence including child sequences can now resume operations.

8.3.3 System Fault Summation

We do not want the equipment running when there is a problem, so we program the logic to have any active fault stop the equipment and prevent it from starting. This is accomplished by inserting the *System Fault Sum*

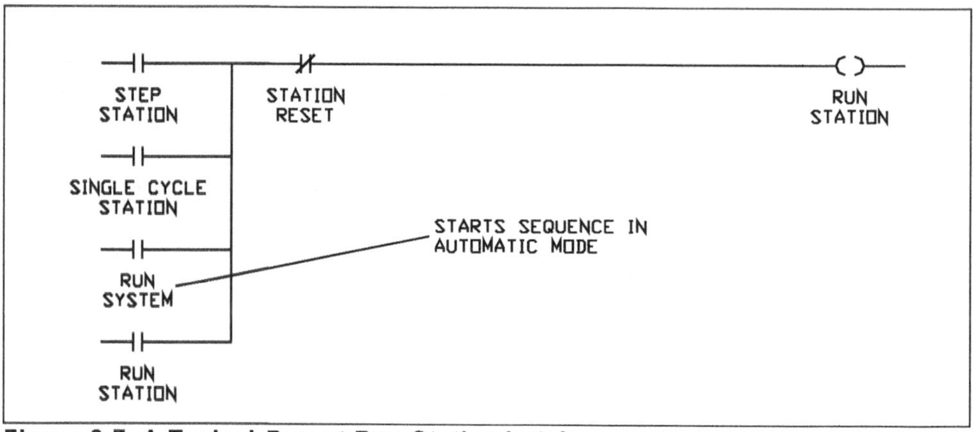

Figure 8-7. A Typical *Parent Run Station* Latch

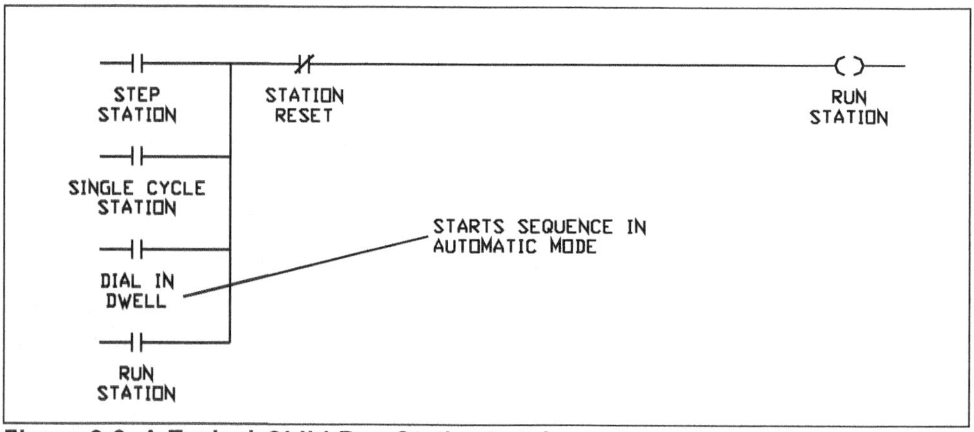

Figure 8-8. A Typical *Child Run Station* Latch

into the *Run System* latch. Each sequence already has a *Station Fault Summation,* so it is just a matter of collecting these and any system-level faults to create the *System Fault Sum,* as shown in Figure 8-9. The system-level faults include *Fault Master is Off, Fault Low Air Pressure,* and *Fault Hydraulic Pump Over Load Tripped.*

A sequence is frozen when it develops trouble and generates a fault. The fault drops the *Station Enable* latch through the *Station Fault Sum* to suspend execution of the sequence. The *Station Fault Sum* also drops the *Run System* latch through the *System Fault Sum.* When *Run System* turns off, none of the remaining sequences will be able to restart but are allowed to finish. Child sequences are also unable to restart because the parent sequence cannot trigger them when it is not executing. The overall result of a station fault is to stop the faulted station immediately and cycle stop the remaining stations. This places the nonfaulted stations into stable

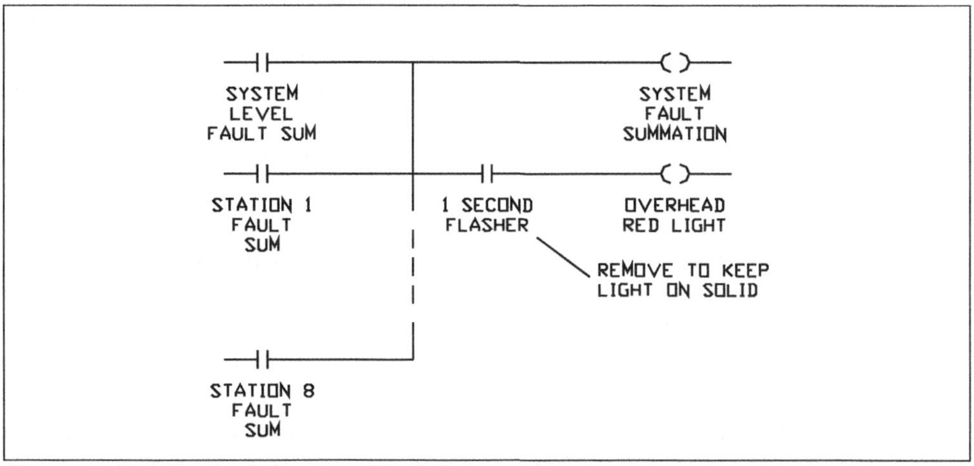

Figure 8-9. A Typical *System Fault Summation*

states and allows entry into the equipment without the loss of motive power messing them up.

8.3.4 Fault Reset

All the fault latches are reset with the *Fault Reset* bit, providing the faulting conditions are gone. The *Fault Reset* bit is signaled directly from the *Fault Reset* push buttons, as shown in Figure 8-10. The *Fault Reset* push buttons are not used directly in the fault latches for several reasons. One reason is there may be several *Fault Reset* push buttons and including both would make the fault latches more complicated. The main reason is in case the logic needs to change. Instead of changing 300 fault latches, change only one *Fault Reset* rung.

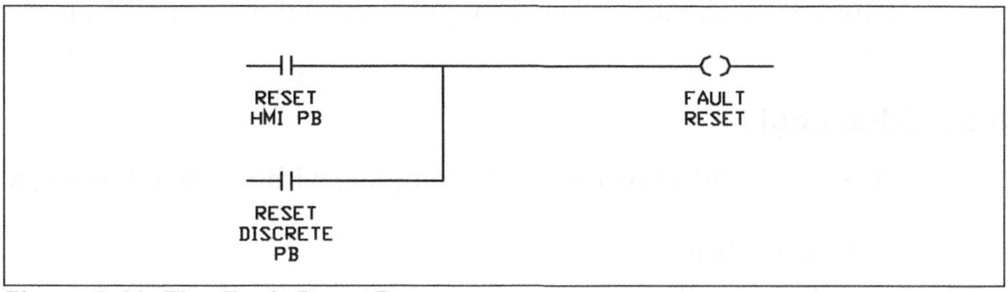

Figure 8-10. The *Fault Reset* Rung

8.3.5 System Ready

Another condition to satisfy before the machine can start in automatic mode is to make sure the station mechanisms are synchronized with their sequences. The home position is the position all the mechanisms need to be in when a sequence starts. A sequence proceeds by commanding its mechanisms to move from the home position, so it is important that the mechanisms be in this position when the sequence starts. By requiring the mechanisms to be in their home position before letting the associated sequence start guarantees that the mechanisms and the sequence are synchronized. Each sequence has a *Station Home Summation* to make it easier to collect this information.

A frozen sequence will most likely not be in its home position. A frozen sequence is a sequence that was already started, as indicated by the *Run Station* latch being on. As a result, the system control knows if a sequence is starting anew or resuming from its current position by using the *Run Station* latch from each station. A *System Ready* rung as shown in Figure 8-11 is constructed to inform the system control when the mechanisms and sequences are synchronized.

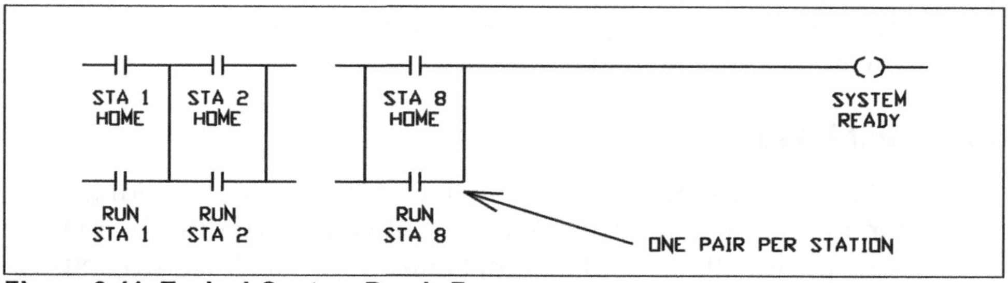

Figure 8-11. Typical *System Ready* Rung

Bypassing the *Station Home Summation* with the *Run Station* bit poses no problems since a started sequence is already synchronized with its mechanisms. If there is a chance the mechanisms of a frozen sequence will move, either by gravity or by hand, then forward checking with position bits should be used. Forward checking is discussed in Chapter 5 on checking sensors.

8.3.6 Other Logic

The system control would not be complete without the following logic:

- Mode Selection
- Machine Running
- In Cycle Pilot Light
- Manual Enable

This logic was discussed in Chapter 7 on step and single-cycle control. It is mentioned here because it is located with the system-control layer in the machine-control program.

8.4 Starting Auxiliary Equipment with the Machine

Auxiliary equipment can also start when the equipment is started with the *Cycle Start* push button, to make the equipment easier to operate. The *Run System* latch is then used to energize them in automatic mode. The operator still has to start the auxiliary equipment separately in the other three operation modes, so they do not have to run when working on a station that does not require them. The auxiliary equipment, such as a hydraulic pump, tends to be noisy, so why run them when you do not need them?

When auxiliary equipment is started with the equipment, and not beforehand, then it cannot be included in the *System Enable* rung or the machine will not be able to start. Instead, the information that the auxiliary equipment is running and that there are no problems with it, is inserted in each *Station Enable* latch as needed. If Station 2 is the only station that requires the hydraulic pump, then its *Station Enable* rung will be the only enable rung to include information about the hydraulic pump, as shown in Figure 8-12.

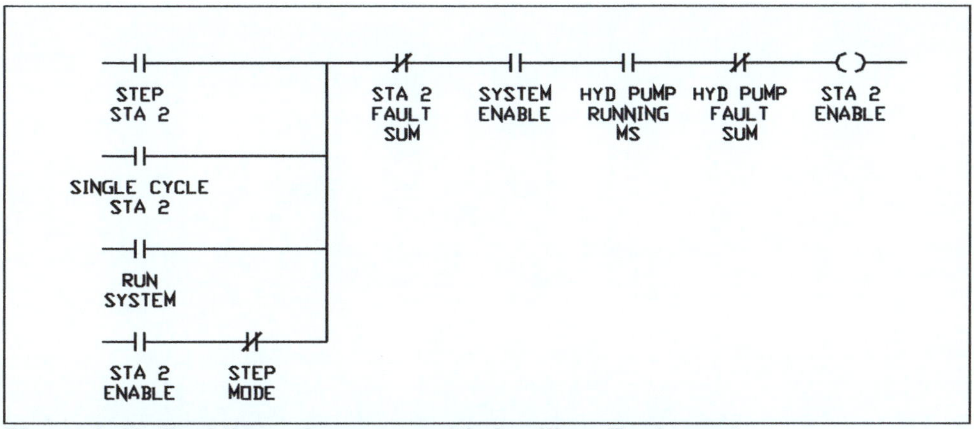

Figure 8-12. Enabling a Sequence with Auxiliary Equipment

All of the problems with the hydraulic pump can be summed up in a *Hydraulic Pump Fault Sum,* to make the information easier to use and keep the *Station Enable* rungs simpler.

8.5 Program Structure

Program structure is how all the logic is organized within the machine-control program, and it generally follows the form of the machine with the logic for Station 1 coming before the logic for Station 2, and so on. The exception is the location of parent sequences. A parent sequence has to be placed before its children for logic scan reasons. The parent sequence contains the shift register and this logic needs to be evaluated before the child sequences are triggered to start. There will be more on this in Chapter 9 on parent/child sequences. Independent sequences can be placed anywhere after the system layer, although they are shown after the child sequences in Figure 8-13.

8.6 Summary

The bulk of the system control is the operational control of the equipment in automatic mode, with the *System Enable* and the *Run System* signals providing coordination. The sequences still do all the work and keep their autonomy. Their independence allows a faulted sequence to stop immediately, while letting the remaining stations cycle stop, making it easier for the operator to restart the equipment.

Overhead
First Scan
Always On
Duty Timers
Low Battery Warning

System Layer
Fault Reset
Master Is Off Fault & Delay
Low Air Pressure Fault & Delay
Auxiliary Equipment Faults
System Level Fault Sum
System Fault Sum
Overhead Red Stack Light
Mode Selection
System Enable
Manual Enable
System Ready
Run System
System Running
In Cycle Pilot Light
Overhead Green Stack Light
System Warning Sum
Overhead Yellow Stack Light

Parent Sequence
Station Home Sum
Station Independent Enable
Step Station
Single Cycle Station
Station Enable
Run Station
Motion Objects
Station Reset
Station Fault Sum
Station Running
Station Output Logic

Child Sequences
Station Bits and Latches

Independent Sequences
Station Bits and Latches

Miscellaneous Logic
Message Control Logic
Production Statistics

Figure 8-13. General Structure of a Machine-Control Program

9 Parent/Child Sequences

9.0 Introduction

This chapter will explain how a parent sequence starts and controls its children. A parent sequence has to do more than just start its children; it also has to make sure they start, handle the data they work with, and make sure they are out of the way when the parent gets around to doing some real work. A dial machine will be used to illustrate the parent/child relationship because it is a common form of industrial equipment and a perfect example of a machine that uses parent and child sequences.

This chapter will first explain what a dial machine is, present flow charts for typical parent and child sequences, and then examine the key elements of each sequence. The main thrust of the chapter is the communication, or interlocking, that is required between the parent and its children. The sequences and interlocking presented in this chapter are a useful template for controlling many machines besides dial machines, such as machines with walking beams and indexing buckets.

9.1 The Dial Machine

A dial machine is used where multiple processes are needed to build up a product. The first process loads the base component part and successive processes increase the value of the part by adding more components to it and perform functions like gluing, drilling, and testing. Processes at the end unload the finished product. A dial machine makes for a compact machine because the stations that perform the processes are arranged in a circular pattern around the mechanism that transports the product between the stations. The transport mechanism is a circular plate called a dial, hence the name of the machine. The dial plate has nests equally spaced around its perimeter, and the nests are used to hold and locate the product to be processed by the stations. The machine operates by rotating the dial to move the product from one station to the next. The stations operate by engaging the product beneath it after the product stops moving. The dial rotates again after all the stations are finished, and the cycle of transferring and processing repeats as long as the system control is on. A typical dial machine is depicted in Figures 9-1 and 9-2.

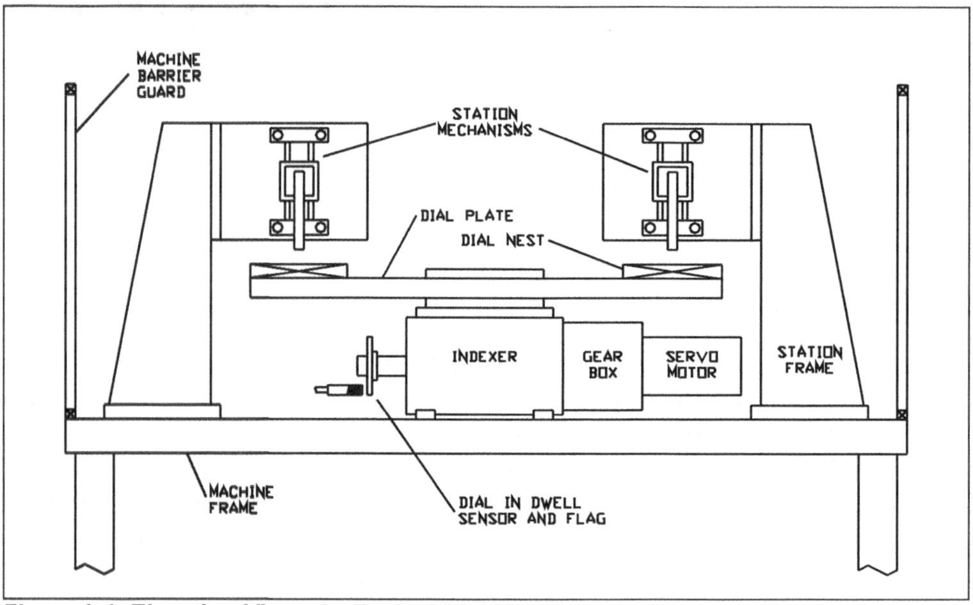

Figure 9-1. Elevation View of a Typical Dial Machine

On a dial machine, the system control handles starting and stopping the machine as a whole; a parent sequence is used to rotate the dial; and child sequences control the stations. A parent sequence is used to control the dial because it is common to all the stations. It knows when the stations can operate and knows when the dial can rotate. The stations cannot operate when the dial is rotating, and the dial cannot rotate when the stations are engaged with the product; so the dial sequence coordinates all this activity.

Dial machines tend to work with smaller product or with less demanding processes to keep the diameter of the dial small or the spacing between the stations close. A typical dial plate is 60 inches or less in diameter. As the diameter of the dial increases, its inertia increases dramatically by the fourth power of its radius, making it harder to start and stop the dial. Gravity also causes the dial plate to sag at its perimeter, making it harder for the stations to find the parts in the nests. The dial machine can be stretched out into a straight line to work with large parts or with processes

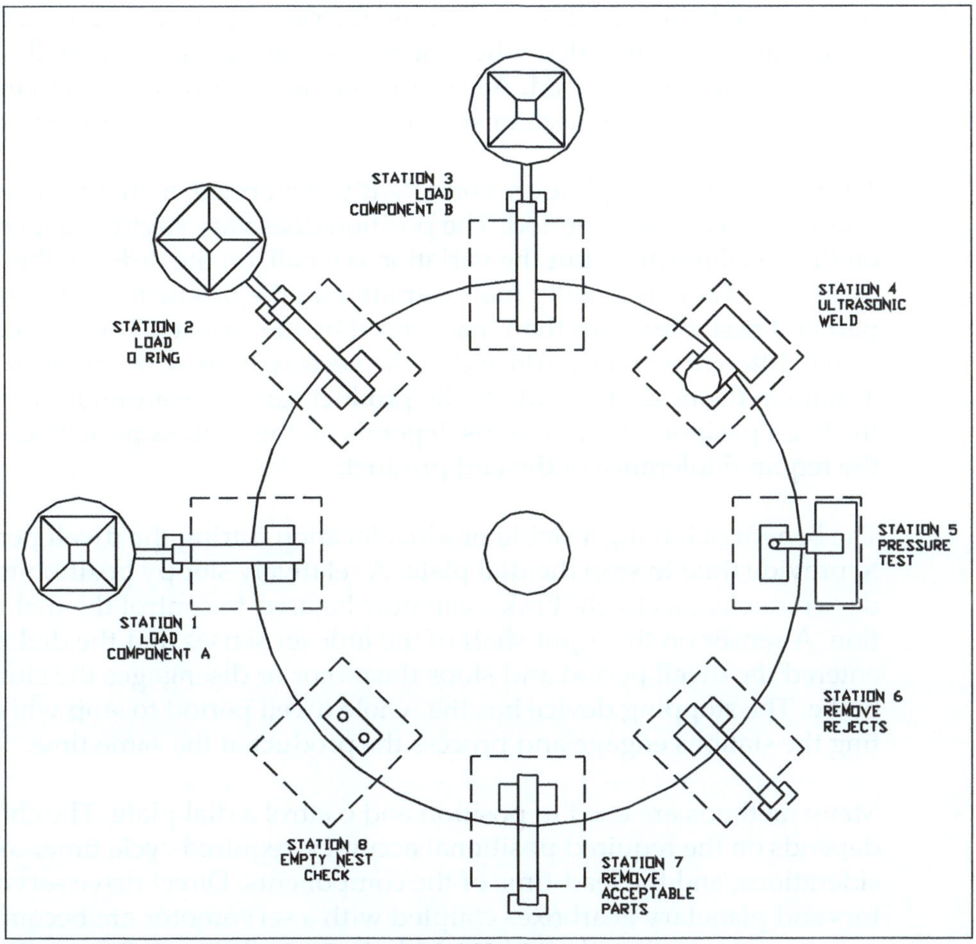

Figure 9-2. Top View of a Typical Dial Machine

that require lots of space. The dial plate is then replaced with a walking beam or another mechanism to facility the new arrangement. Nevertheless the machine control still follows the same format as the dial machine.

9.1.1 The Dial

The most common method used to rotate the dial plate is a cam-style indexer driven by a motor. The indexer has gears, cams, and lobes inside it to create an intermittent rotation on its output for a continuous rotation of its input. The output of the indexer is attached to the dial and provides a dwell period, where the dial is stationary, and an index period, where the dial is rotating. A typical indexer is set up with a dwell period of 90° of input shaft rotation and an index period of 270° of input shaft rotation. A motor is connected to the input shaft of the indexer through a clutch/brake and a gearbox. The gearbox provides the mechanical advantage, and the clutch/brake is used to start and stop the dial's rotation. A clutch/

brake is used with AC motors to allow the motor to run continuously. The clutch/brake is eliminated when the motor is started and stopped, such as when a servomotor is used. An overload clutch is always used to minimize damage when a mechanism entangles itself with the rotating dial.

The nests on the dial plate are consistently positioned, no matter where the indexer is in its dwell period. The position does vary slightly, depending on the dwell position, but the variation is small, on the order of thousands of an inch. There is usually more variation in the product location from part and nest tolerances than that caused by the indexer. The slight dimensional difference in the parts and in the nests will cause the parts to move around in the nests. How much the product can move around relative to the fixed position of the stations depends on the process performed and the required tolerance of the end product.

The benefit of having a stable product location during the dwell period is to provide time to stop the dial plate. A relatively sloppy positioning device, such as a clutch/brake, can now be used to control the dial rotation. A sensor on the input shaft of the indexer senses that the dial has entered the dwell period and stops the motor or disengages the clutch/brake. The stopping device has the whole dwell period to stop while letting the stations engage and process the product at the same time.

Many methods are used to position and control a dial plate. The choice depends on the required positional accuracy, required cycle time, cost considerations, and the lead-time of the components. Direct drive servomotors and planetary gearboxes coupled with a servomotor are becoming more popular to directly replace the indexer, as the cost of these devices decrease and their accuracy increases. The clutch/brake can be eliminated when the motor is started and stopped with the use of a brake, a regenerative DC drive, dynamic braking of a DC motor, DC injection with an AC motor, or a servomotor. Servomotors have the advantage of smoother and faster starting and stopping and can position the dial to the same spot within the dwell period for increase positional accuracy.

9.1.2 Dial In Dwell Sensor

No matter how the dial is positioned and controlled, there will always be a sensor to tell when the dial is in the dwell period. This sensor is responsible for timing when the stations can operate, for stopping the dial, and to pull mechanisms clear of the dial if the dial fails to stop. The encoders on servomotors are not used for these purposes because the overload clutch can separate the motor from the dial. Inaccuracies in the indexer can cause a buildup of positional error over many cycles, and a servomotor needs a homing sensor anyhow. Figure 9-3 shows the *Dial in*

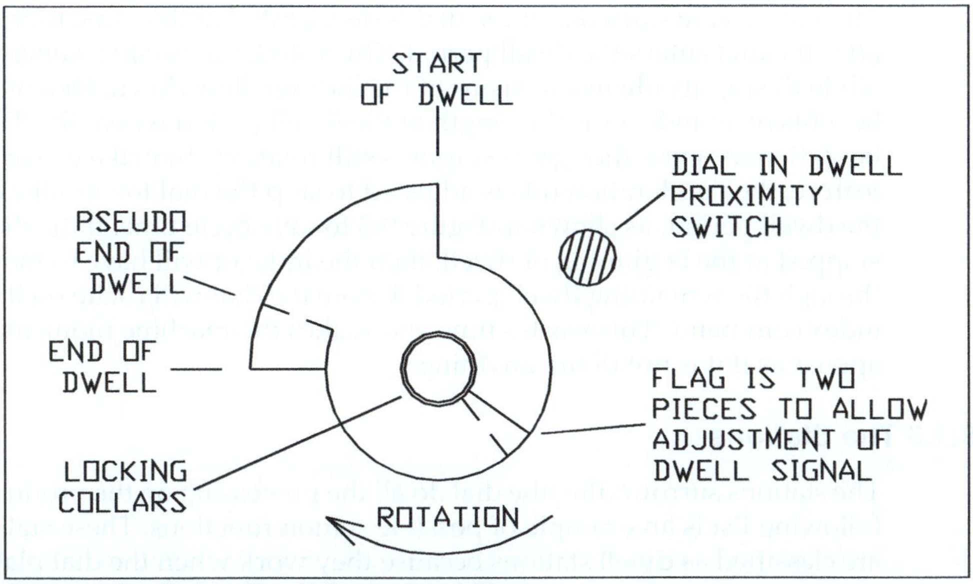

Figure 9-3a. Dial in Dwell Sensor and Flag

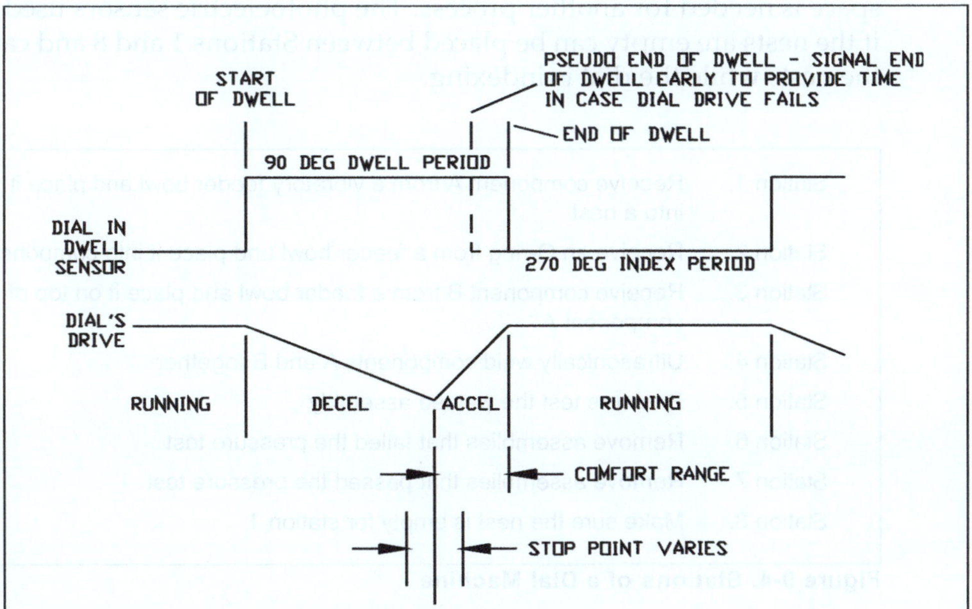

Figure 9-3b. Dial in Dwell Sensor Timing Diagram

Dwell sensor, the flag used to trigger it, and timing considerations. The sensor energizes the input to the programmable logic controller when the dial is in the dwell period as a fail-safe feature. If the sensor fails or a wire breaks, then the logic will respond as if the dial failed to stop.

The stations are started and the dial is decelerated at the same time, just after the dial enters the dwell period. This is to let the stations operate while the input side mechanisms of the indexer slow down. One of the benefits of an indexer is the length of the dwell period where the dial will be stationary, even though the input is still rotating. To make use of this feature, the deceleration rate is adjusted to stop the dial toward the end of the dwell period, as shown in Figure 9-3 to save cycle time. If the dial was stopped at the beginning of dwell, then the indexer will have to move through the remaining dwell period, before the dial will rotate on the next index command. This wastes time and makes the machine momentarily appear as if it is not doing anything.

9.1.3 The Stations

The stations surrounding the dial do all the processing on the product. The following list is an example of possible station functions. These stations are classified as dwell stations because they work when the dial plate is stationary. Stations that operate on the product while the dial is indexing are called *indexing stations*. Station 8 could be an indexing station if the space is needed for another process. The photoelectric sensors used to see if the nests are empty can be placed between Stations 1 and 8 and can scan the nests while the dial is indexing.

Station 1.	Receive component A from a vibratory feeder bowl and place it into a nest
Station 2.	Receive an O-ring from a feeder bowl and place it into component A
Station 3.	Receive component B from a feeder bowl and place it on top of component A
Station 4.	Ultrasonically weld components A and B together
Station 5.	Pressure test the welded assembly
Station 6.	Remove assemblies that failed the pressure test
Station 7.	Remove assemblies that passed the pressure test
Station 8.	Make sure the nest is empty for station 1

Figure 9-4. Stations of a Dial Machine

9.2 The Dial Sequence

The flow chart for a simple dial sequence is shown in Figure 9-5. This sequence will be modified later to add more features, but for now let's examine its basic structure.

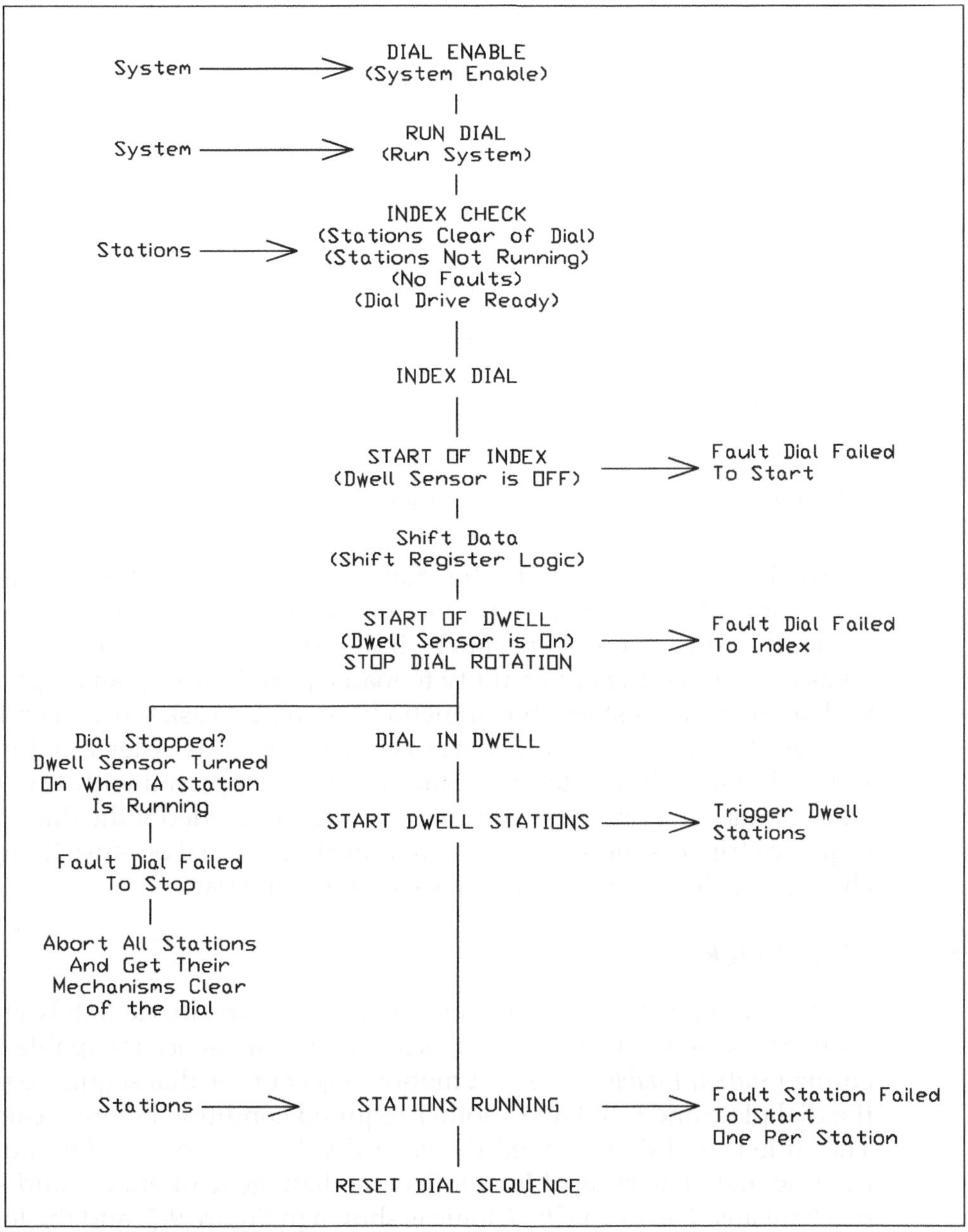

Figure 9-5. Simple Dial Sequence

9.2.1 Starting the Dial Sequence

The dial sequence will start after the operator presses the *Cycle Start* push button, provided the machine is ready to operate, capable of operating, and safe to operate. *System Enable* and *Run System* will check all the required conditions and turn on the dial's *Dial Enable* and *Run Dial* latches. The control logic without the single-cycle and stepping features is shown in Figure 9-6.

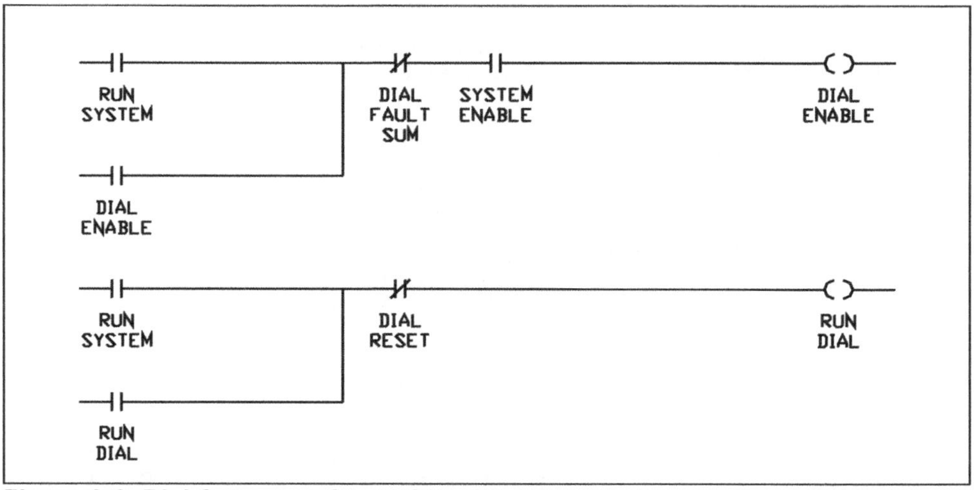

Figure 9-6. Dial Sequence Control Logic

The dial sequence first rotates the dial and then triggers the stations. The disadvantage of this execution order occurs when the machine is first started, causing an empty nest to be indexed into Station 2 because Station 1 has not yet had the opportunity to load a part. Indexing an empty nest will not occur on restarts, but somebody is going to ask why, when it does happen. The sequence could just as easily reverse this order, but the dial is indexed first to show later how Station 1 can be replaced with a manual load/unload station. Extra shift register logic is needed if the dial sequence triggers the stations before indexing the dial because there is a chance a station could operate twice on the same part.

9.2.2 Index Check

The first motion object of a sequence normally starts immediately after the sequence is started (*Run Station*), provided the sequence is capable of executing (*Station Enable*). The first motion object of the dial sequence rotates the dial plate and will have another required condition before it can start. The added condition is called the *Index Check* and is included because a rotating dial plate is capable of inflecting damage to operators and station mechanisms. The *Index Check* logic is shown in Figure 9-7, and the logic for the first motion object is shown in Figure 9-8.

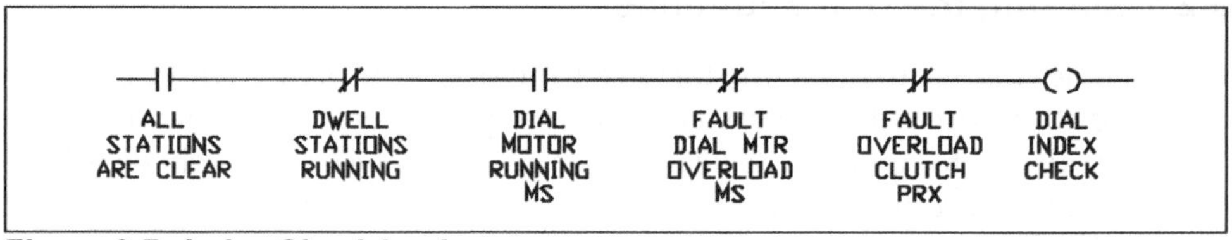

Figure 9-7. *Index Check* Logic

The *Index Check* includes the following conditions:

- **All Stations Clear.** One way of making sure the dial plate will not shear off mechanisms is to check the mechanism's positions. An *All Stations Clear* rung is constructed by summing up a *Station Clear of Dial* rung from each station. The station sequences already have a *Station Home* summation that could be used instead of constructing a separate rung. A separate rung is needed when the station's home position does not coincide with the clear position.

Most often, however, a separate rung is needed because the required cycle time does not permit a station the luxury of returning to its home position before the dial is needed to rotate. An example is Station 3, which receives Component B from a feeder bowl and places it on top of Component A. The dial needs to rotate as soon as the station places Component B and pulls away from the dial because the pick/place mechanism is so long. The pick/place mechanism retracts back to the feeder bowl while the dial is rotating.

- **Dwell Stations Running.** The station mechanisms may be clear of the dial now, but an executing sequence could fire them into a rotating dial plate. The dial plate is prevented from indexing if any station sequence is running, to make sure this does not happen. The station sequences are not executing when the machine is first started because they are started after the dial is rotated. However, on the next machine cycle, the stations will be executing. Notice in Figure 9-5 that the dial sequence resets and starts over after the stations are triggered. A station sequence could also be executing on a machine restart because it was previously frozen due to a fault. The nonfaulted stations were permitted to cycle stop. The construction of the *Dwell Stations Running* rung is similar to the *Machine Running* summation, except that it includes only the critical stations.

If a sequence needs to operate while the dial is rotating, as is the case with Station 3, then the *Station Running* bit is replaced with logic to inform the dial that the station sequence is beyond the point where it engages the dial. See the section below on modifying the *Station Running* bit.

- **Dial Drive Ready.** The logic has to make sure the dial will rotate once commanded because the dial plate has such potential to harm and cause damage. We do not want the dial plate to move unexpectedly and create an ambush condition, as was discussed with the *Manual Enable* bit in Chapter 3 on output logic. As a result, the logic has to check all the conditions that could prevent it from rotating. In the construction of the *Index Check* rung (Figure 9-7), a contact from the dial motor's motor starter is included to make sure it is running. The faults used to monitor the motor's overload relay and the dial's overload clutch are also included.

The motor starter and fault information is redundant when the dial's prime movers are started with the system control. Motive power, including the motor running signal, is checked by the *Dial Enable* bit via the *System Enable* bit. The motion object that commands the dial plate to rotate will not start if the *Dial Enable* bit is off (Figure 9-8). The faults will also turn off the *Dial Enable* bit via the dial sequence's fault summation, *Dial Fault Sum*, so the motor's faults do not need to be included in the index check. The dial's drive information is included in the index check for two reasons. The first is for redundancy—by including dial drive information, we know it is checked and people looking at the code will know it is checked. The second reason is to show what to include in the index check when the component parts are manually loaded and unloaded, as will be discussed.

9.2.3 Rotating the Dial

The motion object in Figure 9-8 starts and stops the dial rotation and monitors the dial's drive. The motion initiation latch, *Index Dial*, turns on when all the required conditions are met and sends a signal to the output logic to engage the clutch. This motion object has two motion complete latches, *Start of Index* and *Start of Dwell*. The *Start of Index* latch will wait for the *Dial in Dwell* sensor to turn off, and then the *Start of Dwell* latch will wait for the sensor to turn back on. Without these two motion complete rungs, the start of the dwell condition will be detected immediately because the *Dial in Dwell* sensor is already on when the dial is commanded to index (Figure 9-3).

The two motion complete rungs also provide reference points in the logic. The *Start of Index* latch and associated o/s is used to advance the shift register and start stations, which operate while the dial rotates (Indexing Stations). The *Start of Dwell* latch and associated o/s are used to stop dial rotation and to stop the indexing stations.

Using two motion complete rungs instead of two complete motion objects prevents the *Dial Enable* from freezing the sequence between starting rotation and stopping it. Stopping rotation takes precedence and should not be inhibited.

Two motion fault latches are also included in the *Index Dial* motion object, since there are two motion complete rungs. The separation of the faults provides more information to assist in tracking down problems. Depend-

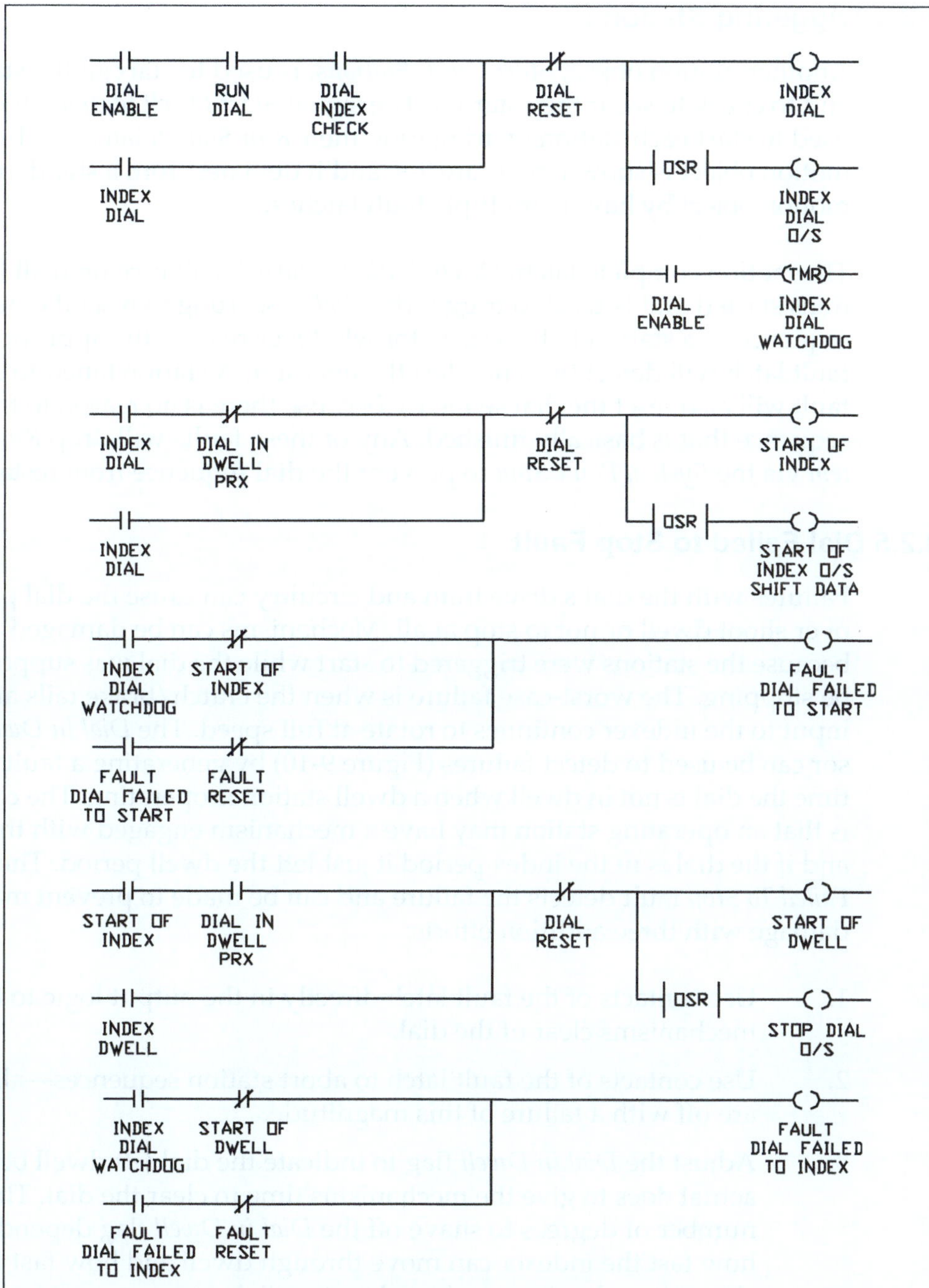

Figure 9-8. First Motion Object of the Dial Sequence

ing on which fault was tripped, we can determine whether the problem is the output, clutch/brake, or sensor.

9.2.4 Triggering Stations

Another motion object, *Start Dwell Stations,* is used to start all the stations and to check to see if they started. The signal *Start Dwell Station O/S* is used to start each station by triggering their *Run Station* latches. This motion object is shown in Figure 9-9, and it deviates from a standard motion object by having multiple fault latches.

The motion complete latch, *Dwell Stations Started,* will turn on if all the stations started and is used to trigger the *Dial Reset* rung to reset the dial sequence. If a station fails to start, for whatever reason, the appropriate fault latch will detect this and alert the operator. A station failed to start fault will also reset the dial sequence because there is no reason to freeze a sequence that is basically finished. Any of these faults will drop *Run System* via the *System Fault Sum* to prevent the dial sequence from restarting.

9.2.5 Dial Failed to Stop Fault

Failures with the dial's drive train and circuitry can cause the dial plate to over shoot dwell or not to stop at all. Mechanisms can be damaged because the stations were triggered to start while the dial was supposed to be stopping. The worst-case failure is when the clutch/brake fails and the input to the indexer continues to rotate at full speed. The *Dial in Dwell* sensor can be used to detect failures (Figure 9-10) by generating a fault anytime the dial is not in dwell when a dwell station is operating. The concept is that an operating station may have a mechanism engaged with the dial, and if the dial is in the index period it just left the dwell period. The *Dial Failed To Stop* fault detects the failure and can be made to prevent machine damage with three addition efforts:

1. Use contacts of the fault latch directly in the output logic to force mechanisms clear of the dial.
2. Use contacts of the fault latch to abort station sequences—all bets are off with a failure of this magnitude.
3. Adjust the *Dial in Dwell* flag to indicate the dial left dwell before it actual does to give the mechanisms time to clear the dial. The number of degrees to shave off the *Dial in Dwell* flag depends on how fast the indexer can move through dwell and how fast the slowest mechanism can just clear the dial.

If a sequence needs to operate while the dial is rotating, as the case with Station 3, the *Station Running* bit in the *Dwell Stations Running* summation is replaced with logic to inform the dial that the station sequence is beyond the point where it engages the dial. See Section 9.4.4 below on modifying the *Station Running* bit.

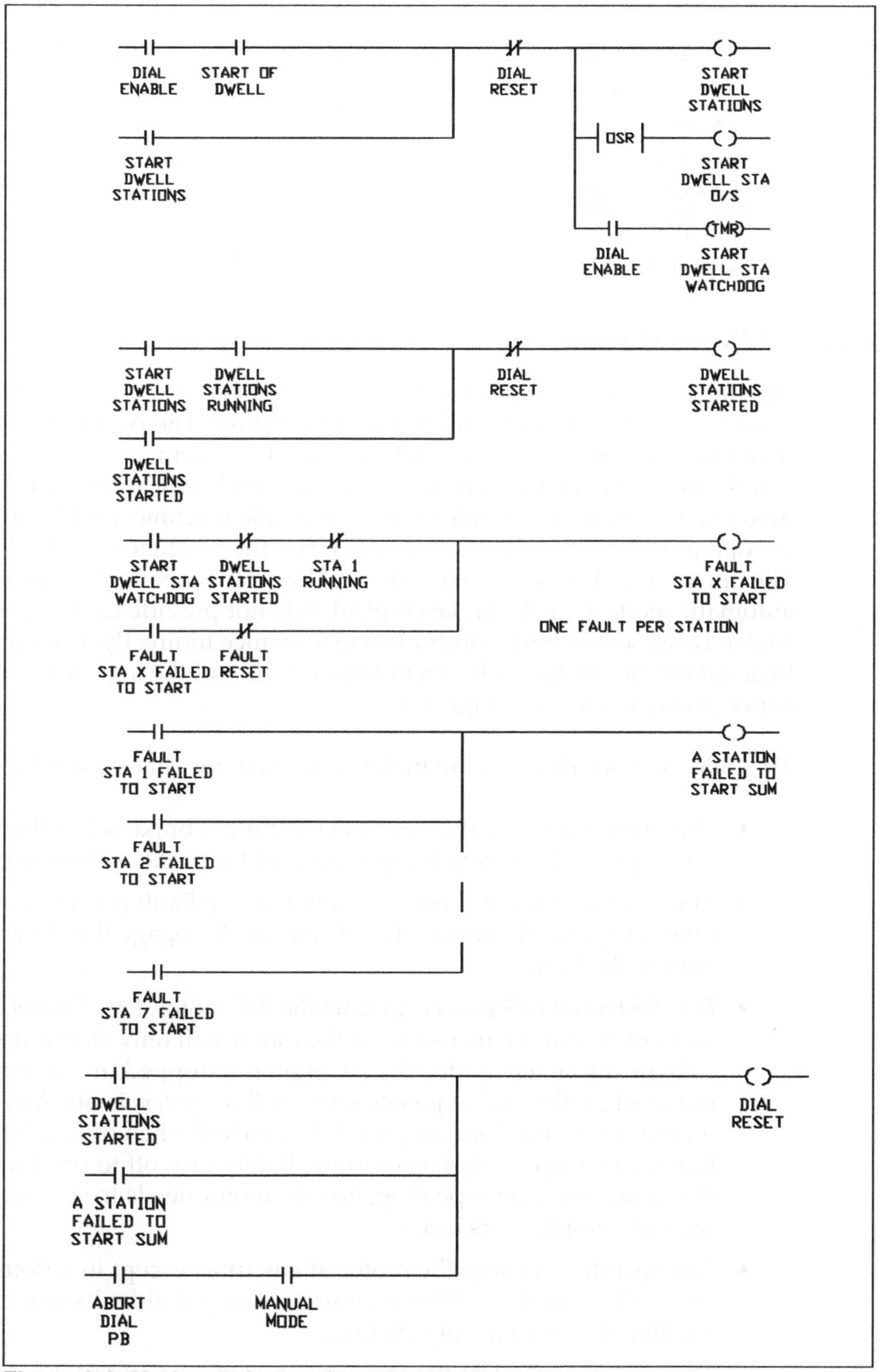

Figure 9-9. Motion Object to Start Station Sequences

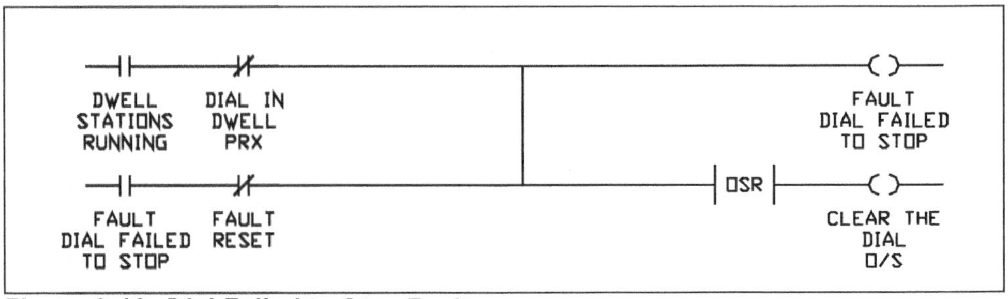

Figure 9-10. ***Dial Failed to Stop*** **Fault**

9.2.6 Dial Prime Mover

The dial sequence assumes the dial's prime mover is a fixed-speed AC motor coupled to the indexer with a clutch/brake. The AC motor is treated as an auxiliary device controlled by the system control. *System Enable* allows it to run, and *Run System* starts and stops the motor. The motor starts when the operator starts the whole machine, and it runs until the operator stops the machine or penetrates the machine's safety devices. The operator will have to start and stop the motor separately in any non-automatic mode. Clutch/brakes typically do not provide feedback, so the *Dial in Dwell* sensor will monitor its performance indirectly. The output logic for the AC motor is shown in Figure 9-11, and the logic for the clutch/brake is shown in Figure 9-12.

The following points cover the major features of the dial's output logic:

- One spring-return valve controls the clutch/brake when the valve is energized, the clutch is engaged, and the brake is disengaged.
- The *System Enable* and the *Dial Failed to Stop* Fault overrides all other output logic to stop the AC motor, disengage the clutch, and engage the brake.
- *Run System on O/S* is used to start the AC motor. *Run System* only turns on in automatic mode, so the motor will only start automatically in automatic mode. The AC motor is stopped in automatic mode when the dial sequence resets if *Run System* is off. An operator may press the *Cycle Stop* push button just when the dial starts to index, causing *Run System* to immediately turn off to prevent the dial sequence from repeating, but the motor needs to stay on until the dial completes its index.
- The operator can stop the motor at any time, except in automatic mode. We do not want the operator creating dial faults when the machine is automatically cycling.
- The operator can start the motor at any time, provided the clutch is disengaged. Motion should not occur just because the motor was

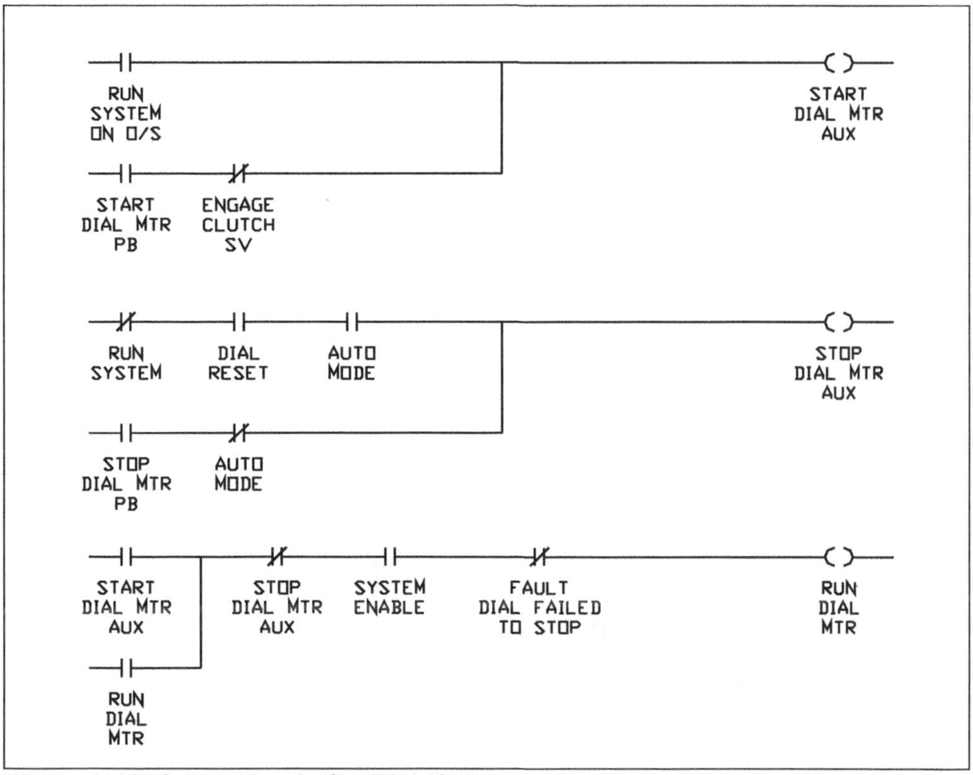

Figure 9-11. Output Logic for Dial Motor

started. The *System Enable* will prevent the motor from starting if it is not safe to do so. Commanding the dial to rotate in manual mode is a two-step process. First start the motor, then command the clutch. Running the motor by itself makes it easier to debug the motor without having to worry about the dial hitting something.

- The *Dial in Dwell on O/S* will always trigger the clutch to disengage. The *Dial in Dwell on O/S* is a one-shot signal that is generated when the dial just enters dwell. The trigger from the dial sequence *Stop Dial O/S* in the auxiliary rung is therefore redundant. This feature was added to allow the dial to be commanded to index in manual mode and have it stop, just like it would in automatic mode, to make it easy to setup and check the *Dial in Dwell* sensor and flag. The dial can be commanded to index, then the location of the flag/sensor is checked. The sensor and flag is adjusted as needed and after several trials the process is done.
- The dial is jogged by engaging the clutch when the *Jog* push button is pressed and by disengaging the clutch when the *Jog* push button is released. The jogging feature does not work very well unless the motor speed can be decreased; otherwise, the motion is too abrupt.

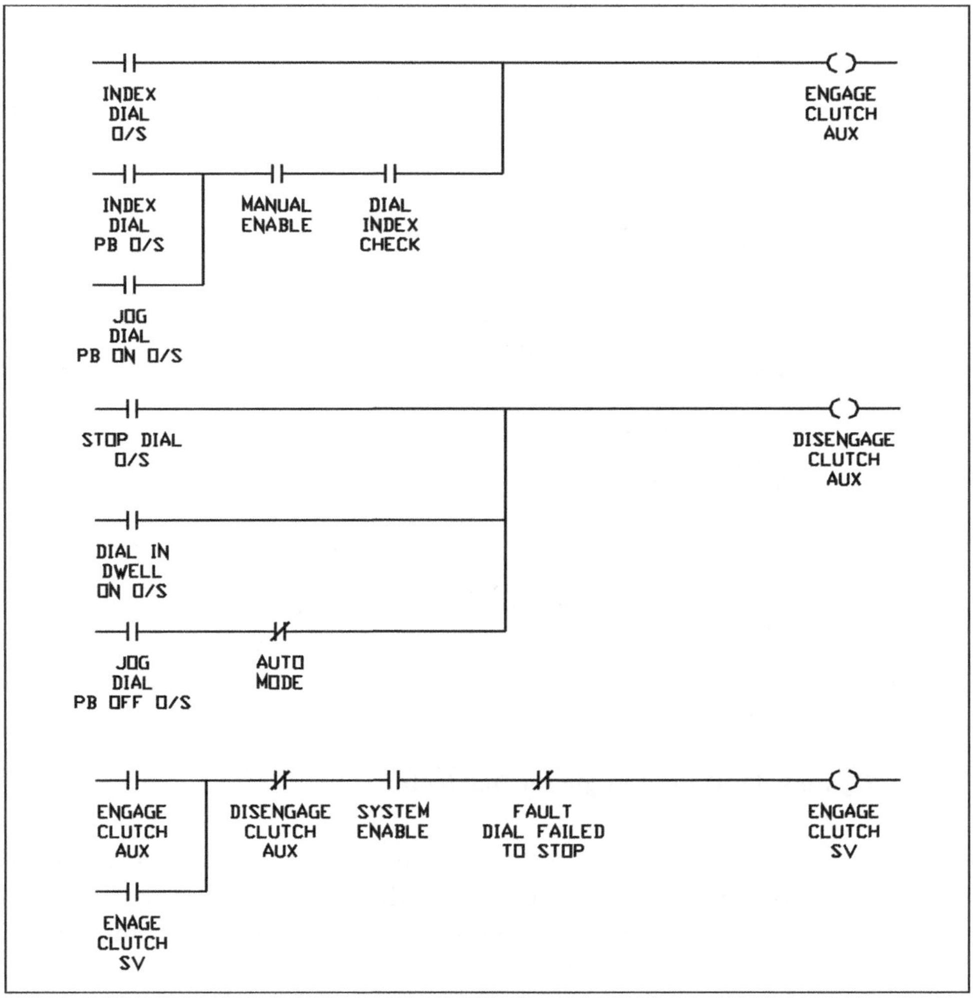

Figure 9-12. Output Logic for Dial Clutch/Brake

9.2.7 Shifting Data

Tracking the product through the machine is the responsibility of the shift register. Simple shift registers use only one bit for each possible product position within the machine. The single bit remembers if the product is either good or bad. The shift register bit for the first position, Station 1, is set to indicate product was loaded. The dial sequence then moves this data to shift register position 2 when the dial rotates the product to Station 2. Station 2 looks at shift register Position 2 to determine if it should process the product or not and updates the shift register, if it successfully finished. Stations will only process the product to increase the product's value when the corresponding shift register bit is set to indicate the product is worth the additional value. The nest beneath the station could be empty or the product failed a prior process when the corresponding shift register bit is off.

More complex shift registers use multiple bits, words, and even multiple words for each possible product position. The additional information is used to record measured values, to differentiate if the product is bad versus the nest is empty, and to log the specific reason why a part was rejected, say with an error code. Additional information helps in troubleshooting the machine and in enabling the machine to make better decisions. One way the shift register can assist in diagnosing problems is to track the nest number with the measured values. The correlation between measured values versus nest number will indicate if a nest needs realignment. Machines that produce pharmaceutics, medical devices, or life-safety components, such as collapsible steering wheel columns, will need data collected and archived for possible future reference.

If large amounts of data are collected, it is usually easier to shift pointers to the data tables rather than whole files because one word is shifted instead of many. Data pointers are implemented in programmable logic controllers through the use of indirect addressing. More information on indirect addressing and shift register logic can be found in the Appendix.

Since the dial sequence indexes the product ahead to the next station, it is a good place to locate the shift register logic. The location of the shift register logic in relation to the stations that use and modify the data is very important because of how the programmable logic controller scans the logic. The dial logic is located above the station logic, so the dial logic is executed first. This lets the shift register advance the data and have it ready before the stations need it. Take the scenario where the execution order of the logic is dial logic, station logic, and then shift register logic. The dial logic will be scanned to trigger the shift register to advance, but the stations execute their logic before the data is actually advanced. The general rule of thumb is to place the shift register logic immediately after the logic that triggers the data to advance.

A good time to shift data is when the dial just starts to rotate (leaving dwell), not necessarily when the dial is told to index. The dial drive could fault between the time the dial is told to index and when it actually starts to rotate. As long as product is physically at a station, the shift register should match the condition of the product to minimize any confusion when looking at the machine.

One problem with shift registers is data integrity. The following list highlights instances where the shift register data could be compromised and the steps used to maintain the data.

- The dial is indexed in manual mode. Use an off one shot from the *Dial in Dwell* sensor to shift the data in manual mode.

- The dial is jogged forward in manual mode, and there is no jog backward feature. Use an off one shot from the *Dial in Dwell* sensor to shift the data in manual mode.
- The dial is jogged backwards in manual mode. Declare all bets are off and set every position in the shift register to reject.
- The programmable logic controller is initialized. Use the first scan bit to set every position in the shift register to reject.
- Servo is homed. Servomotors (the motor, servo drive, and/or motion controller, depending on how the electronics are packaged) have to be "homed" before they can execute a move. Homing is required after the servomotor powers up and after certain servo faults. Commands to the servomotor to initiate a homing routine or signals from the servomotor can be used to set every position in the shift register to reject.
- Loss of motive power. When motive power is lost (emergency stop and/or loss of air pressure), who knows what condition the product was left in? The *Master Is Off* and the *Low Air Pressure* faults can be used to set every position in the shift register to reject.

9.3 Typical Station Sequence

The flow chart for a typical dwell station sequence is shown in Figure 9-13. The flow chart shows the logic that is common for each station. This common logic controls the sequence and interfaces with the dial sequence. Most of the common sequence control logic such as the *Station Enable*, *Run Station*, and the *Station Reset* has already been discussed. New to the station sequences is the logic needed to check the shift register. This logic decides if a station will work on the product or not. The common logic needs to be determined only once then applied to each station, thus saving us program development time. Also shown in the flow chart is the logic unique to each station. This unique logic is depicted as the line item "Station Motion Objects" in Figure 9-13. The unique logic stems from the process the station is required to perform and is created by replacing the flow chart of the station's activities with motion objects. Let's look at the sequence control and the shift register logic.

9.3.1 Station Sequence Control

The dial sequence always starts each station sequence, even if the station is not going to process the product beneath it. The station sequence will operate its mechanisms if the shift register says the product is good; otherwise, the station sequence skips to the end, resets, then waits for the next trigger from the dial. The dial starts each station by triggering each station's *Run Station* latch as shown in Figure 9-14. The dial sequence then

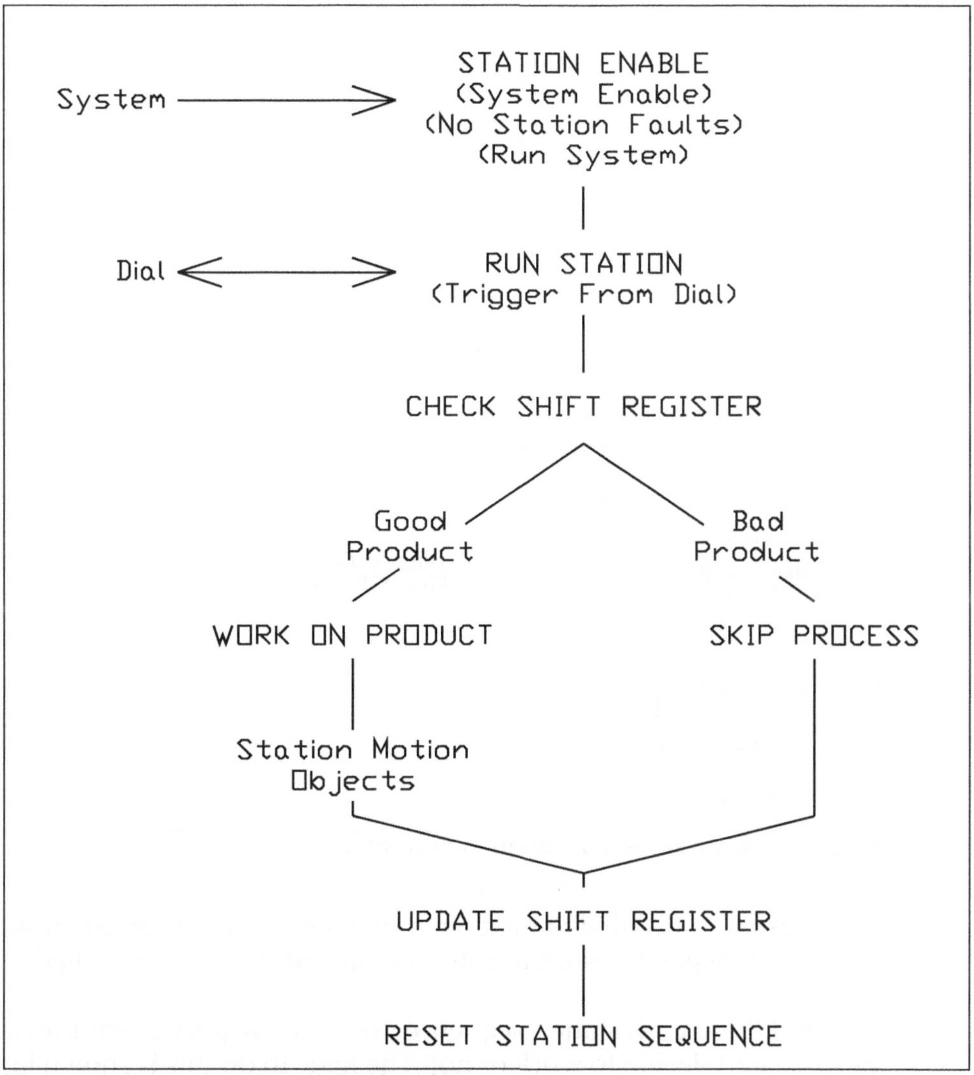

Figure 9-13. Typical Dwell Station Sequence

checks to make sure the stations started by monitoring all the *Run Station* latches. A station fault or perhaps other conditions placed in the station's *Run Station* latch will prevent the station sequence from starting, but at least the "station failed to start" faults in the dial sequence will detect this.

9.3.2 Station Shift Register Logic

A station decides to add value to the product beneath it by looking at the data stored in the shift register corresponding to the station's position. If the product entering the station is good, then the station processes it. If the product is bad, then the station usually skips its operations, but it could do

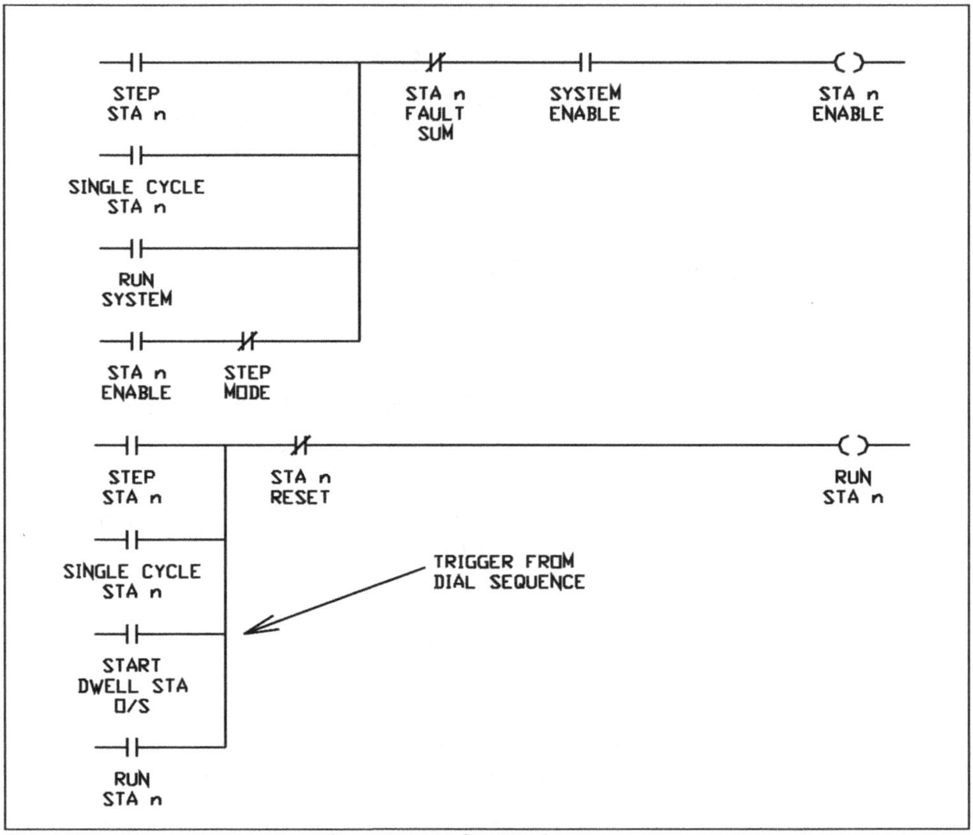

Figure 9-14. Dwell Station Sequence Control

other tasks. Notice that Station 6 removes only rejected product, so it works only when the shift register has tagged the product as bad.

The first thing each station sequence does is to inspect its shift register position and decide to work or not. The logic to do this is shown in Figure 9-15 and is generally referred to as *decision logic*. Decision logic is covered in detail in the next chapter but will briefly be discussed here.

The key feature of the decision logic in Figure 9-15 is that only one of the two rungs will turn on. The *Work On Product* latch will turn on if the shift register is good, and the *Skip Process* latch will turn on if the shift register is bad. Regardless of which latch is set, it will inhibit the other latch from setting. This lock-out feature is important because both of the decision latches can start different sets of motion objects. Only one set of motion objects can execute at a time when they control the same set of mechanisms; otherwise, one part of the sequence will pull the station to the left, while the other part will pull it to the right. This results in nothing working, possible machine damage, and the heavy use of swear words. The flow chart in Figure 9-13 shows the station commanding its mechanisms only when the

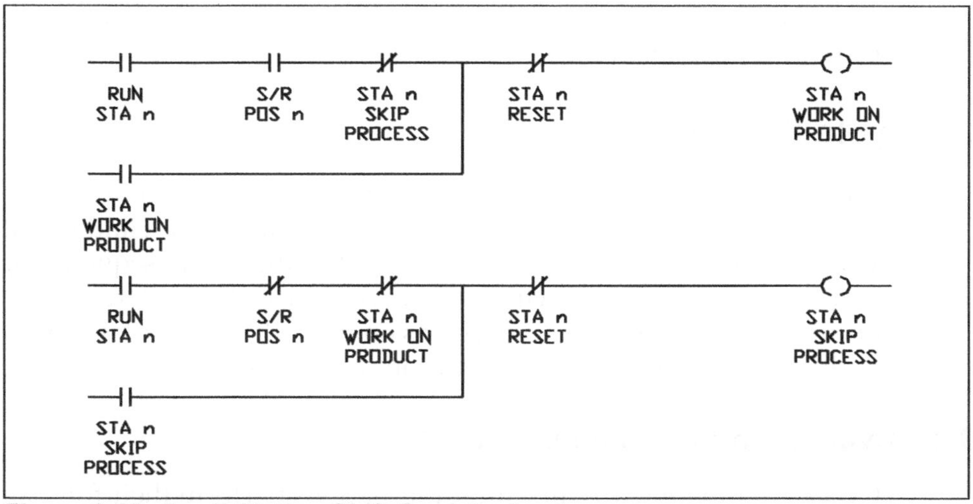

Figure 9-15. ***Dwell Station Shift Register*** **Logic**

Work On Product latch is set. The *Skip Process* latch jumps over the motion objects.

9.3.3 Pseudo Station

Dial drives are sized and set up assuming they will only index once, then stop. When the drive system is allowed to run continuously for more than one index, it is possible for the system to reach a higher speed and gain more momentum. This makes it harder to stop the dial, resulting in longer stopping time and overrunning dwell—the ultimate in dial nightmares. This is not usually a problem with servomotors, such as dial prime movers, because they are commanded to move a specific amount, plus their maximum speeds can be set. The AC motor and clutch/brake system is especially prone to this, as well as any system that relies on a signal to stop.

It is possible to command the dial to rotate for more than one index if the shift register is such that none of the stations will operate. As a result, the dial is told to index again immediately after entering dwell. The dial does not have a chance to slow down and can now accelerate to an even faster speed. Empty shift registers can happen after the controller is initialized or the dial sequence clears the shift register. Normally the product load station prevents dial runaway because the dial has to wait to be loaded but purge modes skips these stations.

A pseudo station can be included to prevent dial runaway. A pseudo station sequence is like any other station sequence, except it always executes (does not bother looking at the shift register) and contains a single timer step. It gives the dial drive time to stop before it is commanded to accelerate again. Programming a pseudo station, if its needed or not, is easy and provides cheap insurance.

9.4 Cycle Stop Points

The cycle stop point is where the machine ends up after the operator presses the *Cycle Stop* push button. With typical parent/child sequences, you can have two cycle stop points, one after the parent sequence operates and another one after the child sequences finish. The difference between one or two cycle stop points depends on if the *Run System* latch is included in the first motion object of the station sequences.

9.4.1 Cycle Stop After Stations Finish

Cycle stopping the machine after the stations perform their function is assumed by the flow charts in Figures 9-5 and 9-13. If the *Cycle Stop* push button is pressed while the dial is indexing, the dial will continue to rotate to the next dwell period, then the stations are told to operate, the machine will cycle stop after the stations finish. If the *Cycle Stop* push button is pressed while the stations are operating, then the machine will cycle stop after the stations finished. This cycle stop point is programmed by not including the *Run System* latch into the first step of the station sequences as shown in Figure 9-16.

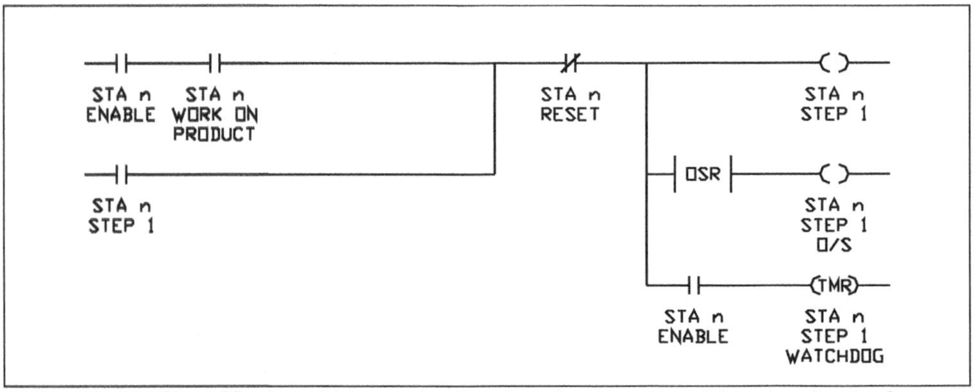

Figure 9-16. First Motion Object of a Station Sequence without a Cycle Stop Point

This method of cycle stopping brings the whole machine to a concluded state with the product below each station completed for the process performed at that station. This eliminates confusion for the operators, especially if the machine stops due to a fault, and the operator must decide the best way to get the machine restarted.

Let's look at a vision system inspection station as an example. This type of station determines if the product beneath it is good or bad. The vision system usually has a monitor displaying the last inspection results. Other indicators included with the operator interface may also display pass/fail status for the station. When looking at these results, it would be best if they matched the product currently beneath the station. Allowing the station to operate during a cycle stop lets the vision system operate and update its results, so they match the product beneath the station. With the other cycle stopping method, where the stations are not allowed to operate after the dial enters dwell, the vision system will yield results from the last machine cycle, thus adding confusion as to the status of the product.

9.4.2 Cycle Stop After Index

The machine can be made to cycle stop after the dial is indexed but before the stations operate. If the *Cycle Stop* push button is pressed while the dial is indexing, the dial will continue to rotate to the next dwell period then the machine will cycle stop. If the *Cycle Stop* push button is pressed while the stations are operating, then the machine will cycle stop after the stations finish. These cycle stop points are programmed by including the *Run System* latch into the first motion object of the station sequences as shown in Figure 9-17.

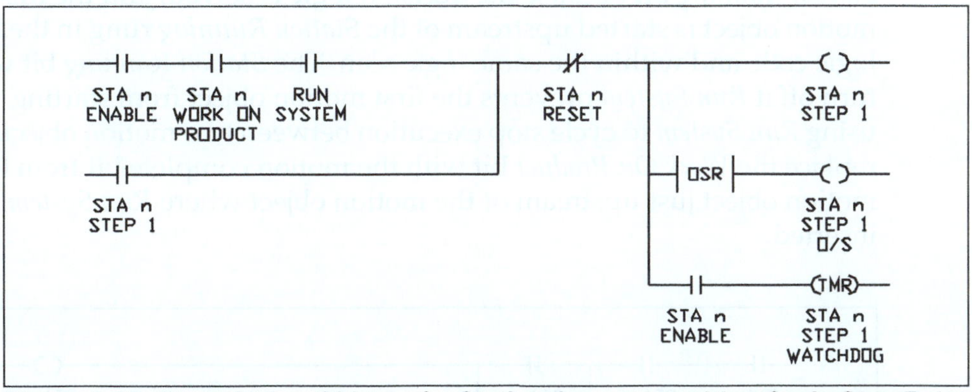

Figure 9-17. First Motion Object of a Station Sequence with a Cycle Stop Point

If the machine was cycle stopped after the dial indexed, the machine knows the station sequences were started because their *Run Station* latches will be on. The station sequences will continue after the machine is restarted when the *Run System* latch turns back on.

Cycle stopping after the dial indexes, without letting the stations operate, brings the machine to rest quicker. It is also a useful debug tool because it allows you to let the stations work on the product, then index the product out of the station for viewing, especially if the station mechanisms hides the product. This is a big help when testing the machine on a station-by-station basis. However, this method can place the product out of sync with product status indicators.

9.4.3 Other Cycle Stop Points

The *Run System* latch can be inserted into any sequence motion object to create a cycle stop point at that step. If the Cycle Stop push button is pressed, the sequence will step along until the step with the *Run System* latch. The sequence will no longer be able to continue until the machine is restarted. Additional cycle stop points can be created as an aid in trouble-shooting.

9.4.4 Modifying the Station Running Bit

The *Station Running* rung will need to be modified when *Run System* is used to interrupt the execution of a sequence. The *Station Running* bit, without modifications, will remain on, even though the sequence is no longer executing because the *Run Station* and *Station Enable* latches will remain on. The necessary modification is shown in Figure 9-18. The modified *Station Running* bit will turn off as soon as the station sequence reaches the decision point and will not turn back on until the first motion object is started. The normal execution of the logic will not even blink the *Station Running* bit because the decision logic is solved, and the first motion object is started upstream of the *Station Running* rung in the ladder logic code and within the same logic scan. The *Station Running* bit will turn off if *Run System* prevents the first motion object from starting. When using *Run System* to cycle stop execution between two motion objects, then replace the *Work On Product* bit with the motion complete bit from the motion object just upstream of the motion object where *Run System* is inserted.

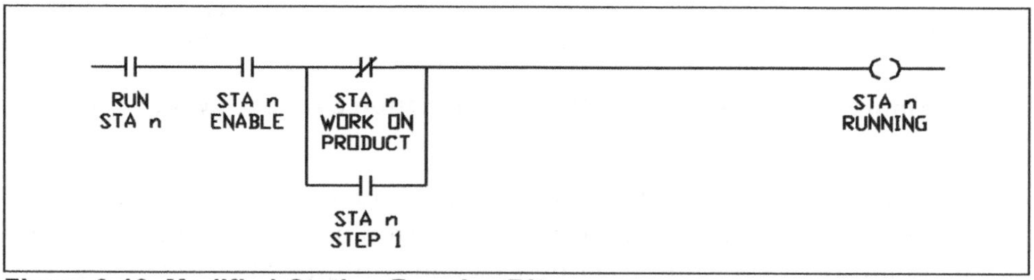

Figure 9-18. Modified *Station Running* Bit

The *Index Check* and the *Dial Failed To Stop* fault both use a *Dwell Stations Running* summation. It was mentioned that if a station sequence needs to operate while the dial is rotating, as the case with Station 3, then the *Station Running* bit is replaced with logic to inform the dial that the sequence is beyond the point where it engages the dial. This logic is very similar to the logic in Figure 9-18. Station 3 is clear of the dial just after it finishes raising after placing Component B. At this point, the dial can rotate without harming the mechanisms of Station 3. The *Station Running* bit for Station 3 is then replaced with a *Station Running On Dial* bit in the *Dwell Stations Running* summation to provide the interlocking for the index check and the failed to stop fault. The construction of this replacement interlock is shown in Figure 9-19.

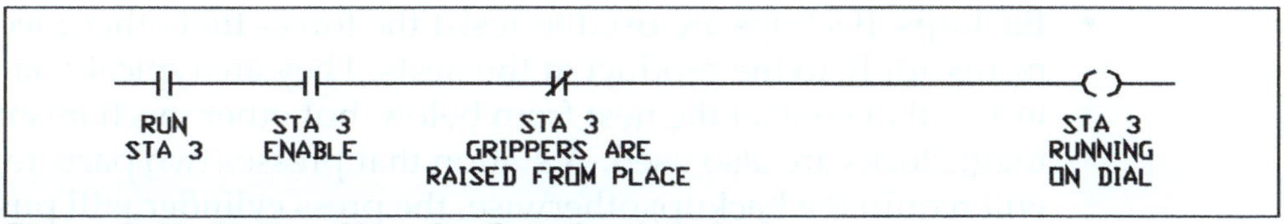

Figure 9-19. Station Clear for Rotation Interlock

9.5 Manual Load/Unload Stations

Many times an operator will load and unload the product from the dial. When this is done, Stations 1, 6, 7, and 8 are no longer needed. To facilitate a manual station on the dial, the following changes are made besides deleting sequences:

- **Operator Safety.** A light curtain is used to protect the operator's safety. Whenever the light curtain is penetrated, the dial is braked to a stop and its prime mover is disabled. The dial needs to stop before the operator can reach any moving object inside the machine. Additional barrier guards are usually needed between the stations to allow the operator to work at the load/unload station while the remaining station command their mechanisms.
- **Palm Buttons.** The *Cycle Start* push button is replaced with palm buttons and a safety rated anti-tiedown relay. The palm buttons start the system control, *Run System*, but the system control does not latch on. This starts the dial's prime mover and the dial sequence. The dial sequence will operate just once without *Run System* latched on, and thus there is no need for a *Cycle Stop* push button.
- **Pass/Fail Indication.** Red and green pilot lights are used to inform the operator if the finished parts are good or bad.

9.6 Dial Mechanisms

The dial sequence can operate other mechanisms besides the dial plate, such as the following devices and mechanisms:

- **Shot Pins.** Shot pins are used to position the dial to an accurate and repeatable position and consist of a tapered pin attached to a cylinder and a mating hole mounted on the dial plate. The pin is inserted into the matching hole after the dial stops rotating. Compliance in the dial drive mechanism allows the dial plate to rotate slightly and align itself with the pin. The shot pins thus locate all the nests horizontally. The dial sequence controls the shot pin because it has to extend before any station can operate and has to retract before the dial can rotate.
- **Backups.** Backups are used to resist the forces that other mechanisms apply to the product in the nests. They are typically large cylinders that contact the nest from below, but other mechanisms like toggle locks are also used. A station that presses two parts together will required a backup; otherwise, the press cylinder will push on the nest, deflect it downwards, and bend the whole dial plate. The backups can be controlled by the station that needs it or from the dial sequence after the dial stops but before the station starts. Backups are also used to slightly move the nest above it to an accurate and repeatable vertical position.
- **Nest Vacuum.** Product can be held in the dial nests with vacuum. The station sequences normally controls this vacuum, so it can be timed with product placement or removal. However, the dial sequence usually controls the nest vacuum when it has to change, regardless of what the station does; some stations may not operate because of the shift register.

To control the above mechanisms from the dial sequence, concentrate on the shot pin, for all the other mechanisms are just variations of it. The shot pin extends after the dial stops rotating but before the stations operate and it has to retract before the dial can rotate. The dial sequence is shown in Figure 9-5, and it will be worth reviewing before proceeding because we will be modifying it for the shot pin.

To extend the shot pin, a new motion object is inserted between the *Dial In Dwell* and the *Start Dwell Stations* steps in the dial sequence. Before, the motion complete rung *Start Of Dwell* was used to start the *Start Dwell Stations* motion object—see the first rung in Figure 9-9. Now, the motion complete rung *Start Of Dwell* will start the *Extend Shot Pin* motion object instead. The Start Dwell Stations motion object is now started after the shot pin is extended.

As it turns out, the best place to retract the shot pins is before the dial is told to index. To retract the shot pins, another motion object is added to the dial sequence between the *Index Check* and the *Index Dial* motion object. The dial sequence does not care if the shot pin is extended or retracted because it is going to retract it anyhow (by default). The shot pins will not retract, just like the dial would not rotate, until all the stations are finished by placing the *Retract Shot Pin* motion object after the *Index Check*.

9.7 Indexing Stations

Indexing stations operate while the parent sequence is transferring the product, and they are generally inspection stations, rather than ones that operate mechanisms. Station 8 can be replace with an indexing station to make room for a new station. The photoelectric sensor that looks for any product remaining in the nest can be moved between Station 8 and 1 and timed to look for product in the nest while the dial rotates the nest beneath it. The empty nest inspection logic can be constructed as in Figure 9-20, assuming only the product can trigger the sensor rather than the nest itself. The motion complete rungs *Start Of Index* and *Start Of Dwell* are used to enable and disable the empty nest check, respectively.

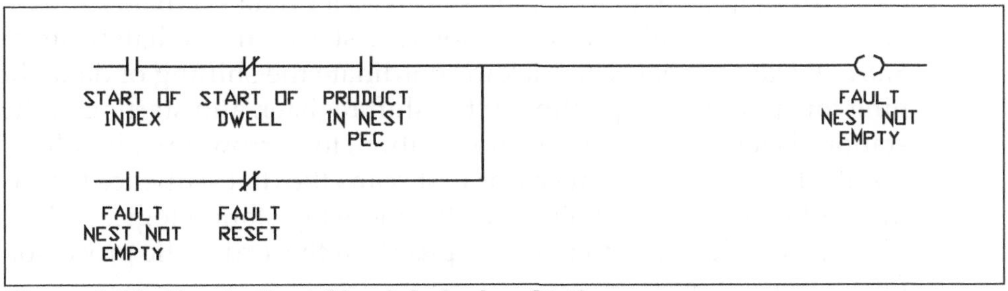

Figure 9-20. Empty Nest Check Indexing Station

A better example of an indexing station is a splice-detection station. Say you have a machine that uses a hitch feed to pull a continuous strip of web through it. Each cycle of the hitch feed will advance the web a fixed and repeatable number of inches (one pitch). A new roll of web is spliced onto the trailing edge of the old roll to keep the machine running, and the splice is made with a tape detectable by a photoelectric sensor. The sensor sees only a small spot but can scan the full pitch length when it looks while the web is moving. The parent sequence for this machine will control the hitch feed and the timing of the sensor. The *Start Of Index* and *Start Of Dwell* signals are used to enable and disable the *Splice Was Found* inspection latch similarly to the way they timed the *Nest Not Empty* fault in Figure 9-20. The splice-detect station logic is shown in Figure 9-21; as can be seen, not much logic is needed. Notice that if the *Splice Was Found* latch is set, it stays on until the start of the next cycle. This provides time to view the latch for troubleshooting and setup.

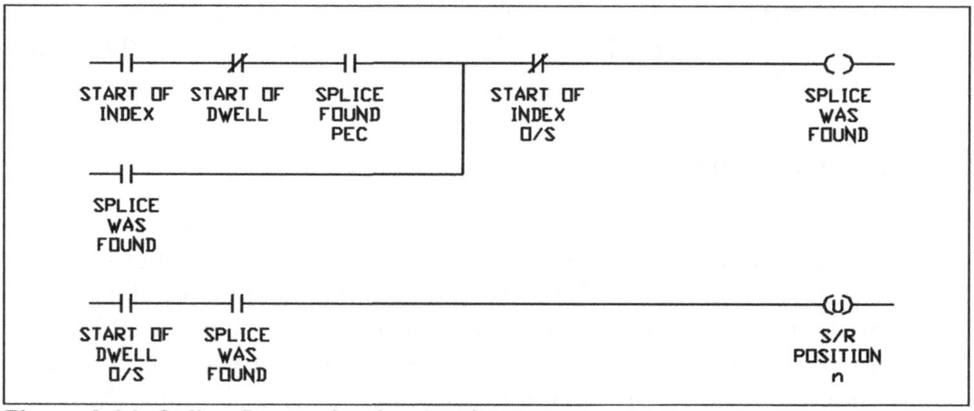

Figure 9-21. Splice-Detect Station Logic

The shift register is updated with the information from the *Splice Was Found* latch on each cycle of the hitch feed and has to be accomplished before the station immediately downstream of the splice station uses the data. To complicate matters, the shift register data is advanced with the *Start Of Index* signal, and the stations are started immediately after the *Start Of Dwell* signal. The trick to coordinate the shifting of data, the scanning for splices, the updating of the shift register, and starting of the dwell stations is location, location, and location, to borrow a reality phase. Locate the logic so the processor first scans the dial sequence that includes the shift register logic, then scans the indexing station logic, and then the dwell station logic. With the logic placed in this order, the processor solves the logic as follows:

Start of index signal
Shift register is advanced
Reset splice was found latch
Enable splice was found latch
...
Start of dwell signal
Disable splice was found latch
Update shift register with splice information
Start dwell stations
...

9.8 Code Generation and Checking

A good programming practice is to first create the flow charts for the parent sequence and one child sequence. Pick the station that loads the parts (usually the first station). Next, use these flow charts to check all the timing and interlocking between the two sequences. The flow charts for the other station sequences can then be quickly written after the interlocking is worked out. The only differences between the station sequences are the actual tasks the stations perform; all the interlocks with the parent sequence are the same.

Playing with the flow charts can check the relationship between parent and child sequences. This type of checking is similar to playing a board game. Tokens can be moved through the flow charts to see how far they go before an interlock stops them. The following is a list of several items to check.

- Cycle stop requested when the dial is indexing.
- Cycle stop requested when the stations are operating.
- Fault occurs in dial sequence before the station sequences are triggered.
- Fault occurs in dial sequence after the station sequences are triggered.
- Fault occurs in a station sequence.
- Restarting after a fault.
- Cycle start requested when the dial is in the dwell position.
- Cycle start requested when the dial is between dwell positions.

10 Decision Logic

10.0 Introduction

Previous chapters discussed how sequences orchestrate machine motion by telling the machine what to do. These sequences execute one step after another following a single path from start to finish. However, there are circumstances where a sequence may need to choose between alternate paths. How a sequence chooses a route is called *decision logic* and will be the focus of this chapter.

10.1 The Pick/Place Example

An example of a multiple path sequence is a pick/place mechanism that includes product inspection. The mechanism has sensors to measure the part or to check for a feature on the part. Its sequence has decision logic to determine what to do, based on the outcome of the inspection. The mechanism will place parts into a dial nest when the inspection is good or dump the part into a reject bin when the inspection is bad. Flow charts for this pick/place mechanism are shown in Figures 10-1 and 10-2. The part is inspected after the grippers are lowered and closed around the part, ensuring the part is correctly orientated for the inspection sensor. Notice that the sequence does not diverge until after the grippers are raised from pick. Raising the grippers from the pick position is common to both good and bad parts, and making the sequence diverge after this step saves us from writing two motion objects for the same move.

10.2 Part Inspection

The logic for the *Inspect Part* step of the pick/place sequence is shown in Figure 10-3. This is an important step as far as program flow is concerned,

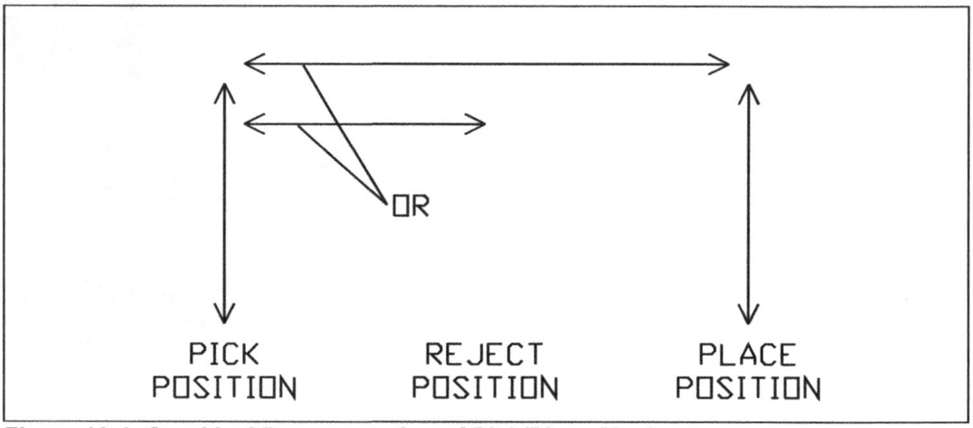

Figure 10-1. Graphical Representation of Pick/Place Mechanism

for it decides which path the sequence will follow, even though the sequence branches after the next step. The four important aspects of decision logic are selection, protection, lockout, and triggering.

10.2.1 Selection

The decision logic determines which path to follow by setting one and only one of the decision latches within the decision group. The appropriate latch is selected by examining the selection criteria. The selection criteria are all the conditions the decision logic uses to make its choice. In the pick/place example, the selection criteria is the presence or absence of a feature on the part, such as a tab. The feature could also be a pneumatic port on the part that is inspected to see if it is bored out or blocked. The inspection can include checking to see if the port is threaded or not.

The example sequence only has two possible outcomes, good or bad, so the decision group only has two decision latches. A typical decision group has one decision latch for each possible sequence branch. All the possible outcomes have to be divided up between the decision latches. For instance, the inputs from two sensors can resolve themselves into four possible outcomes. It is desired to split these possible outcomes between two decision latches, as illustrated by the first two rungs in Figure 10-4. The first decision latch looks to see if only one input is on; if so, the part is good. It does not matter which input is on, only that one of them is on. The second latch then needs to look for the remaining two possible sensor combinations of both inputs on or both inputs off.

10.2.2 Decision Fault

The decision logic can stall a sequence when a possible outcome is overlooked, for no decision latch will set, and thus no sequence branch will

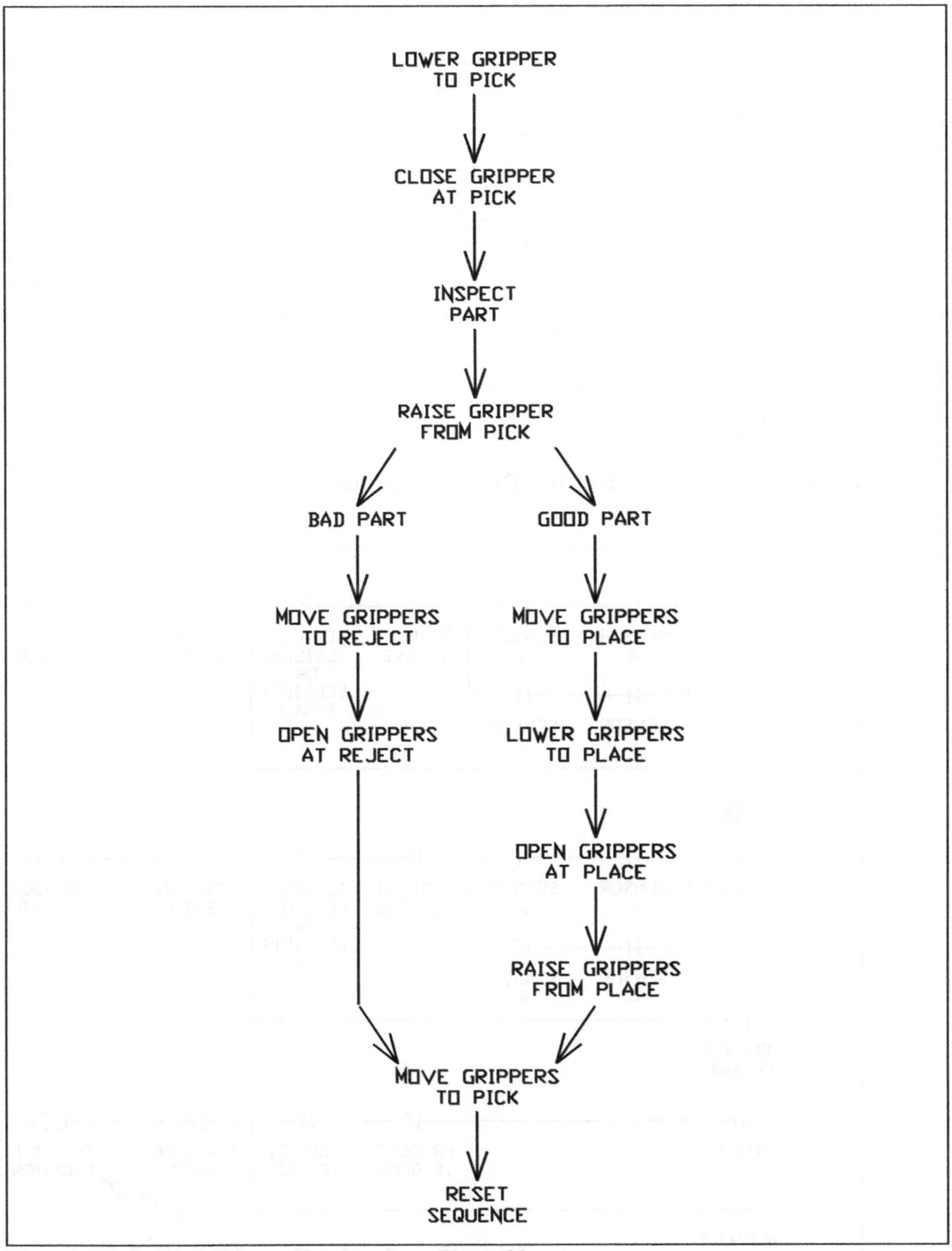

Figure 10-2. Pick/Place Mechanism Flow Chart

execute. With a stalled sequence, the machine will sit idle with no indication of what went wrong.

Three things can be done to prevent decision logic from stalling a sequence. The first is to make sure all possible selection criteria outcomes are accounted for. The second is to program a default decision latch that

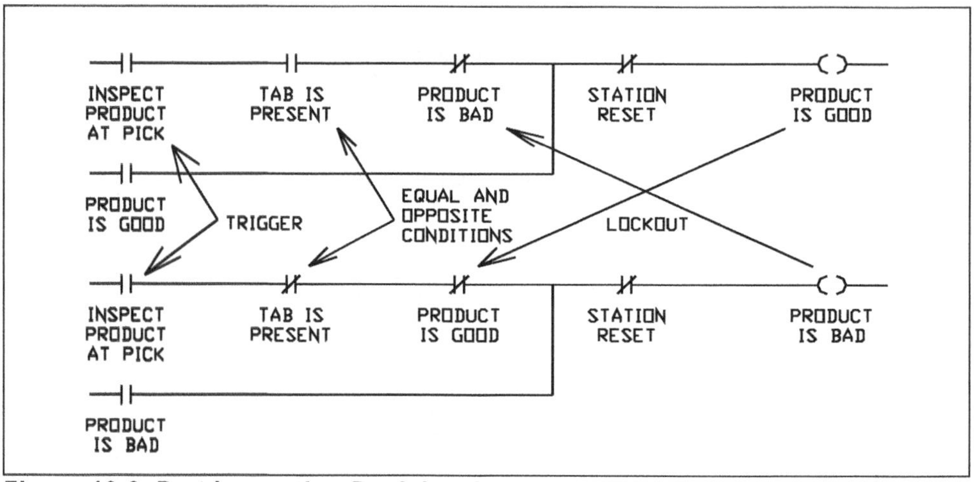

Figure 10-3. Part Inspection Decision Group

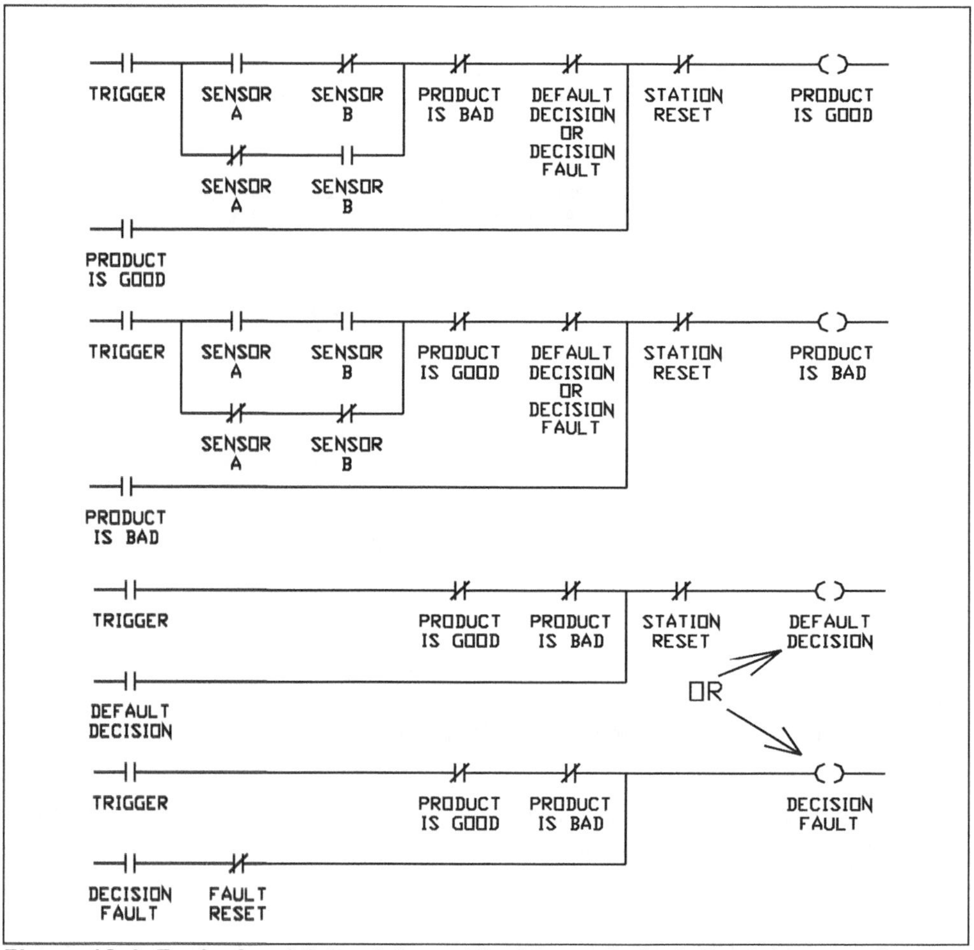

Figure 10-4. Exclusive OR Decision with either a Default Decision or a Decision Fault

will latch on in the event no other decision latch is set. Default decisions usually are programmed for fail-safe operation, such as rejecting the part. The third is to program a decision fault. A decision fault is similar to the default decision latch, in that it will set if no other decision latch is set, but instead of making a default decision, it generates a fault to stop the machine. The theory is that at least you will know when the sequence stalls. The decision fault is mainly used as a safety net when programming with complicated selection criteria. Both a default decision and a decision fault are shown in Figure 10-4. Typically only one type of decision protection is used.

10.2.3 Lock Out

Once the decision is made by setting the appropriate latch, the latch will remain on until the sequence resets. This is to allow tracing of the logical flow through the sequence for troubleshooting. It is also required to lock the decision so it cannot later change its mind. The winning decision latch is used to disable all of the remaining decision latches, preventing them from turning on in the subsequent program scans. A bad decision is one thing, but an unstable decision can cause multiple sequence branches to execute.

10.2.4 Triggering

Triggering is the method of timing when a decision is made. It is important that all the decision latches within a decision group be triggered at the same time and in the same way. The first, and perhaps the most important, reason is to interrogate the inspection sensors at the correct point within the machine cycle. The sensors can be sensing the wrong thing and providing false information at any other point within the machine cycle. By locking the sensors to a specific point within the machine cycle, we ensure the information they provide is valid.

The sensors are also interrogated at other machine cycle points to back check their operation. Looking at the sensors when they are supposed to be in the opposite state allows the machine to check them. Critical sensors should be back checked to ensure they are working properly, for the sensors themselves can fail, debris can falsely trigger them, or the associated mechanisms can fail.

The second reason to trigger the decision latches at the same time is to help ensure that only one decision latch within the decision group is set. The winning latch can then immediately lock out all the losing latches by triggering all the latches at the same time. Lockout is actually performed on the next program scan for all the decision latches logically located above the winning latch location. The decision latches located above the winner were logically solved false for the current program scan; so it is of no importance that the lockout function occurs on the next program scan.

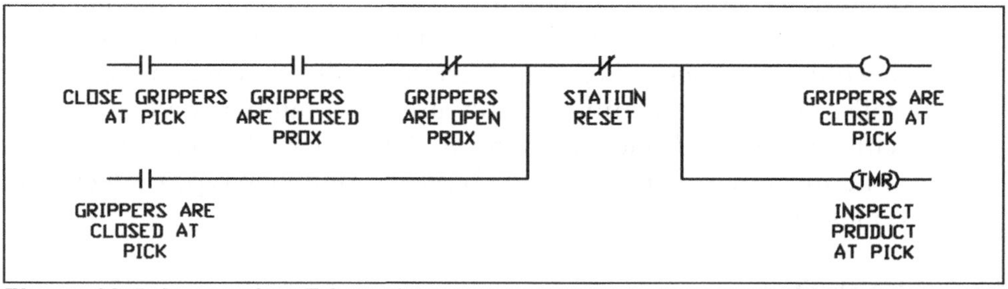

Figure 10-5. Inspection Triggering

Figure 10-5 shows the trigger for the example pick/place mechanism. The trigger *Inspect Product At Pick* occurs a short time after the grippers are closed. The time delay helps to ensure a good reading of the inspection sensor *Tab Is Present*. Time delays are used to account for improperly adjusted sensors and hysteresis in the sensors, and to provide time for the product to settle. The *Grippers Are Closed* proximity switch may turn on too early, and it could turn on while the product is still moving because of the grippers closing on it. The product may also vibrate because of handling and bumping against side rails.

Another method of triggering involves using a once-shot contact in the decision latches. The one-shot contact gives the decision latches only one program scan to make a decision. The trigger shown in Figure 10-5, *Inspect Product At Pick,* turns on at the appropriate time, but it also stays on until the sequence resets. This is a good example of where the lockout function of the decision latches is needed. Without the lockout function, the latch *Product Is Bad* will turn on after the sensor *Tab Is Present* turns off, which could then start executing another branch in the sequence. The use of one-shot triggers prevents this, but I would still program the lockout function to assist in recognizing decision logic when viewing a printout of the code. The lockout function is also required when using either a default decision latch or a decision fault.

10.3 Continuing Sequences

The next step in the pick/place example, after the part is inspected, is to raise the grippers from pick. This is done if the part is good or bad. The logic to activate the motion object for this step is shown in Figure 10-6. The two conditions used to start the motion *Product Is Good* and *Product Is Bad* can be replace with the single condition *Inspect Product At Pick,* which is the inspection trigger. The inspection trigger can also be used to start the next sequence step, since the motion object to raise the grippers is located after the inspection logic in the program, and the inspection occurs within one program scan. The two results of the inspection are shown in Figure 10-6, to illustrate that this step occurs regardless of the outcome of the inspection. It is the preferred method because it makes the code more readable.

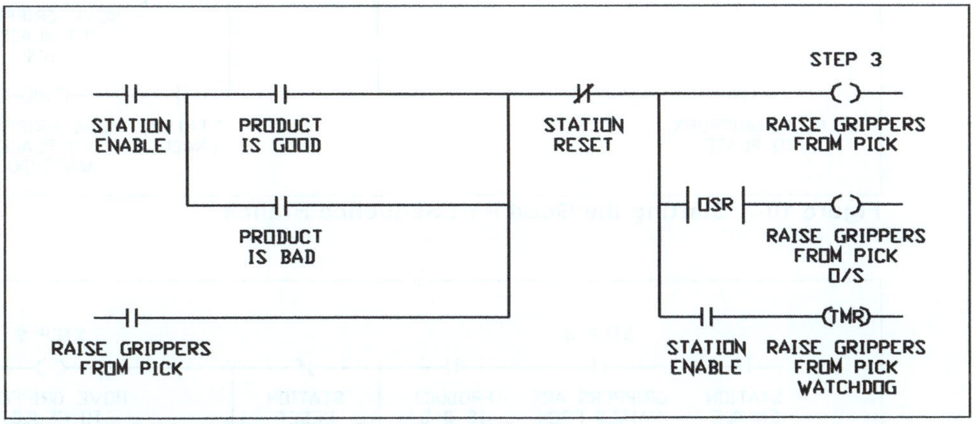

Figure 10-6. Next Step After Part Inspection

10.4 Branching Sequences

The mechanism travels to either the place position to deliver a good part or to the reject position to dump the bad part, after the grippers are raised from the pick position. Since we have two latches to tell us the status of the part, we can use them to select one of two possible sequence branches. We also use the completion of the *Raise Grippers From Pick* motion object to start the appropriate sequence branch. The appropriate sequence branch is started by starting the first motion object in the respective branch. Figure 10-7 shows how the beginning of the good branch is programmed, and Figure 10-8 shows the programming for the bad branch.

10.5 Converging Sequences

Once execution of the appropriate sequence branch is started, the branch will initiate motion, wait for the motion to complete, and then start the next motion until the branch is finished. Sooner or later, the two separate branches have to converge back together. In the pick/place example, the branches merge when it is time to move the mechanism back to the pick position.

When the good branch is finished, the mechanism is positioned up and over the place position with the grippers open. To start the move back to the pick position, we wait until the motion object for step *Raise Grippers*

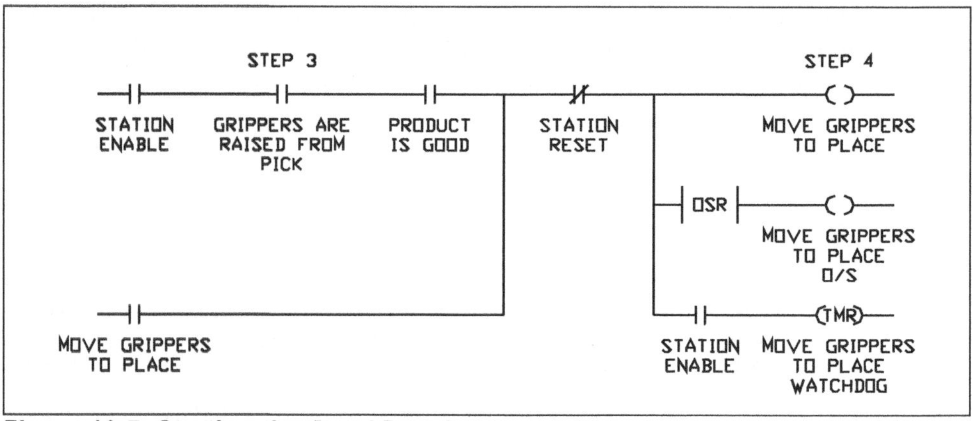

Figure 10-7. Starting the Good Part Sequence Branch

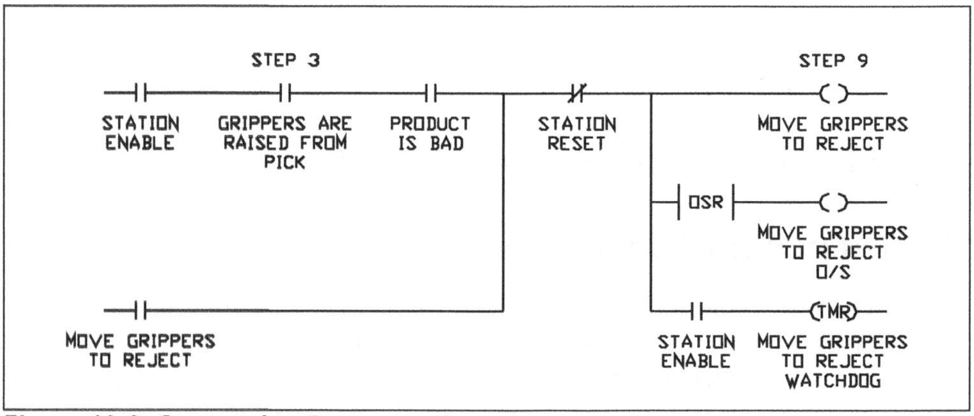

Figure 10-8. Converging Sequence Branches

From Place to complete. When the bad branch is finished, the mechanism is at the reject position raised and with the gripper open. To start the move back to the pick position, we wait until the motion object for the step *Open Grippers At Reject* to complete. After either of these events occurs, the mechanism can return to the pick position. Since the two branches are divergent (only one is selected), then the two events are logically "OR" together as Figure 10-9 shows. If we programmed the decision logic correctly then only one branch will execute.

10.6 Parallel Branches

Another type of sequence branch is one that can execute at the same time another branch is executing. These two branches are executed in parallel. An example of parallel branches is a station that receives product on a pallet. After the station removes the product from the pallet, its sequence splits to further process the product and to get rid of the empty pallet. Additional work on the product can happen at the same time the pallet is

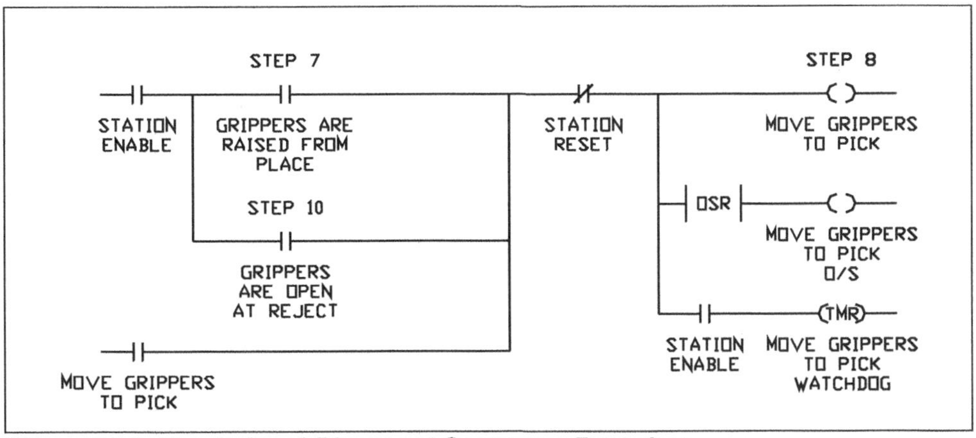

Figure 10-9. Parallel and Divergent Sequence Branches

leaving. Both parallel branches must be finished before the sequence can continue or finish. This is programmed by logically "AND" together the conditions to signify both branches are finished. In general, divergent paths are "OR" together and parallel paths are "AND" together to bring the branches back together as shown in Figure 10-10.

10.7 Conclusion

The decision to load the dial nest or reject the product was based on inspecting the product directly. In Chapter 9 on parent/child sequences, each station on the dial machine made a decision if it should work on the product beneath it or not. The decision was made on the basis of the status of the part as stored in the shift register. As examples, decisions can be based on certain faults, if the system control *Run System* is still on, and if the machine is in purge mode. Decisions are necessary to allow the equipment to handle a variety of circumstances. These circumstances can range from bad parts to resetting a sequence that was waiting for a part when the operator presses the *Cycle Stop* push button. Should the station keep the whole machine running, waiting for a part, or give up and stop?

11 Dependent Sequences

11.0 Introduction

The task of writing programs for machinery is made easier by breaking up the machine into small, manageable sections with each section programmed with its own sequence. The whole machine control is then achieved by using appropriate interlocking between these separate sequences. Dial machines are broken up into separate sequences for each station on the dial and another one for the dial itself. These sequences are then interlocked together in a parent/child relationship. Some machine sections can be programmed for independent operation, such as a vibratory feeder bowl with a part sorter. The bowl sequence runs whenever the system control is on and only concerns itself with keeping the product pipeline to the machine full. There are also sequences dependent on each other. These sequences have appropriate interlocks with each other to transfer parts and so forth. This chapter will focus on how to interlock dependent sequences.

11.1 The Example

A load station on a dial machine with a part inspection substation is a good example of where dependent sequences can be used. The inspection station fetches a part from an input track, inspects it, and then presents only good parts to the load station. Rejected parts are dumped. The load station picks the good parts from the inspection station and places them onto a nest on the dial. A separate sequence is written for the load station and for the inspection station. Figure 11-1 shows the flow chart of the two sequences and the interlocking between them.

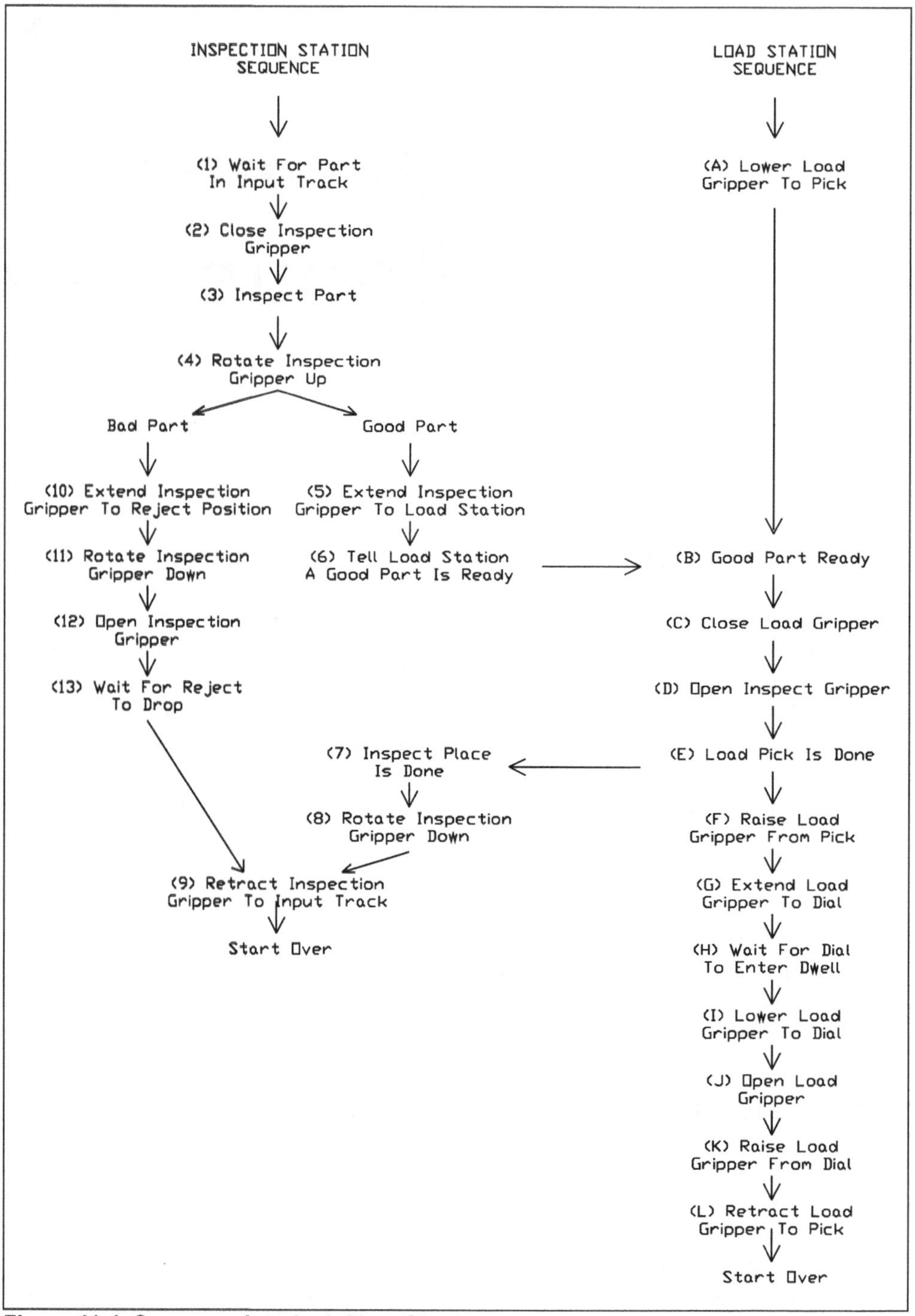

Figure 11-1. Sequence Interlock Flow Chart

The inspection station interlocking logic is shown in Figure 11-2, and the load station interlocking logic is shown in Figure 11-3. As the logic shows, the interlocking uses standard motion objects with extra sequence steps included to make the interlocking cleaner. For instance, the signal from the load station to the inspection station (signal E7) could just have well originated from the *Inspection Gripper Is Open* latch (step D) in Figure 11-3, instead of adding the separate latch *Load Pick Is Done* (step E). Logically, and from a program flow perspective, only one latch is needed. Two separate latches split the functionality of the motion complete rung from the interlocking with the inspection station. The two latches also allow for two separate comments in the program.

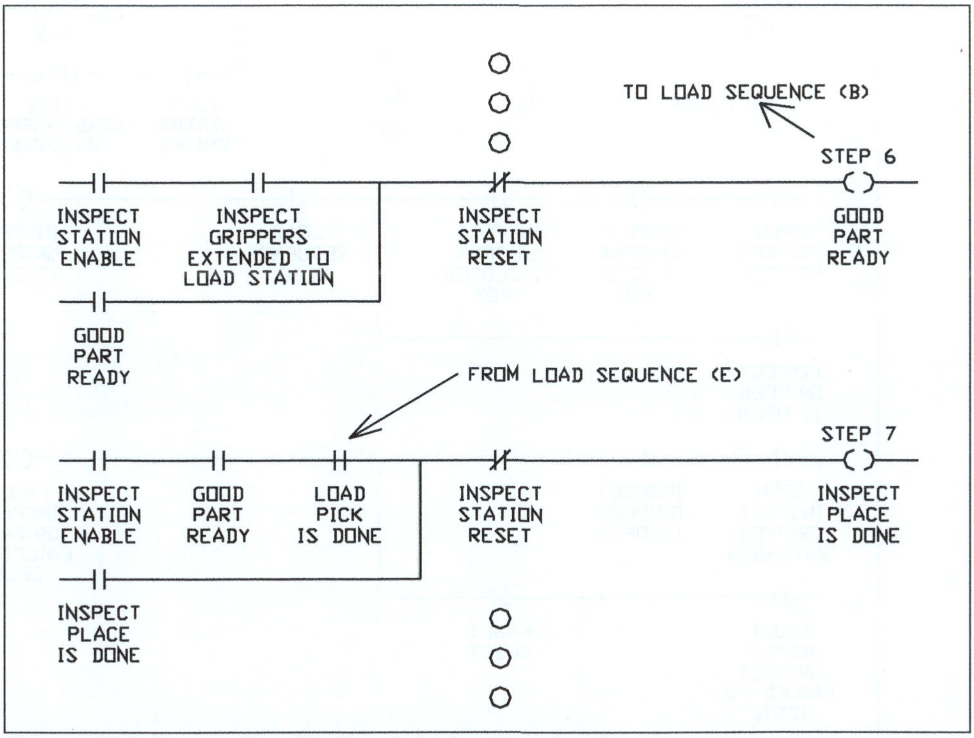

Figure 11-2. Inspection Station Interlocking Logic

11.2 The Problem

One of the best ways to find problems with logic, short of actually running the machine, is to play what-ifs, using the flow charts. You can ask yourself what will this sequence do if the other sequence faults out. For instance, what will the load sequence do if the inspection grippers fail to retract from place (step 8 in Figure 11-1)? The problem is illustrated in Figure 11-4.

In this situation, the inspection station sequence will freeze because of the fault, and all the energized latches will remain on. In the inspection sequence, all the latches up to and including the motion initiation rung of

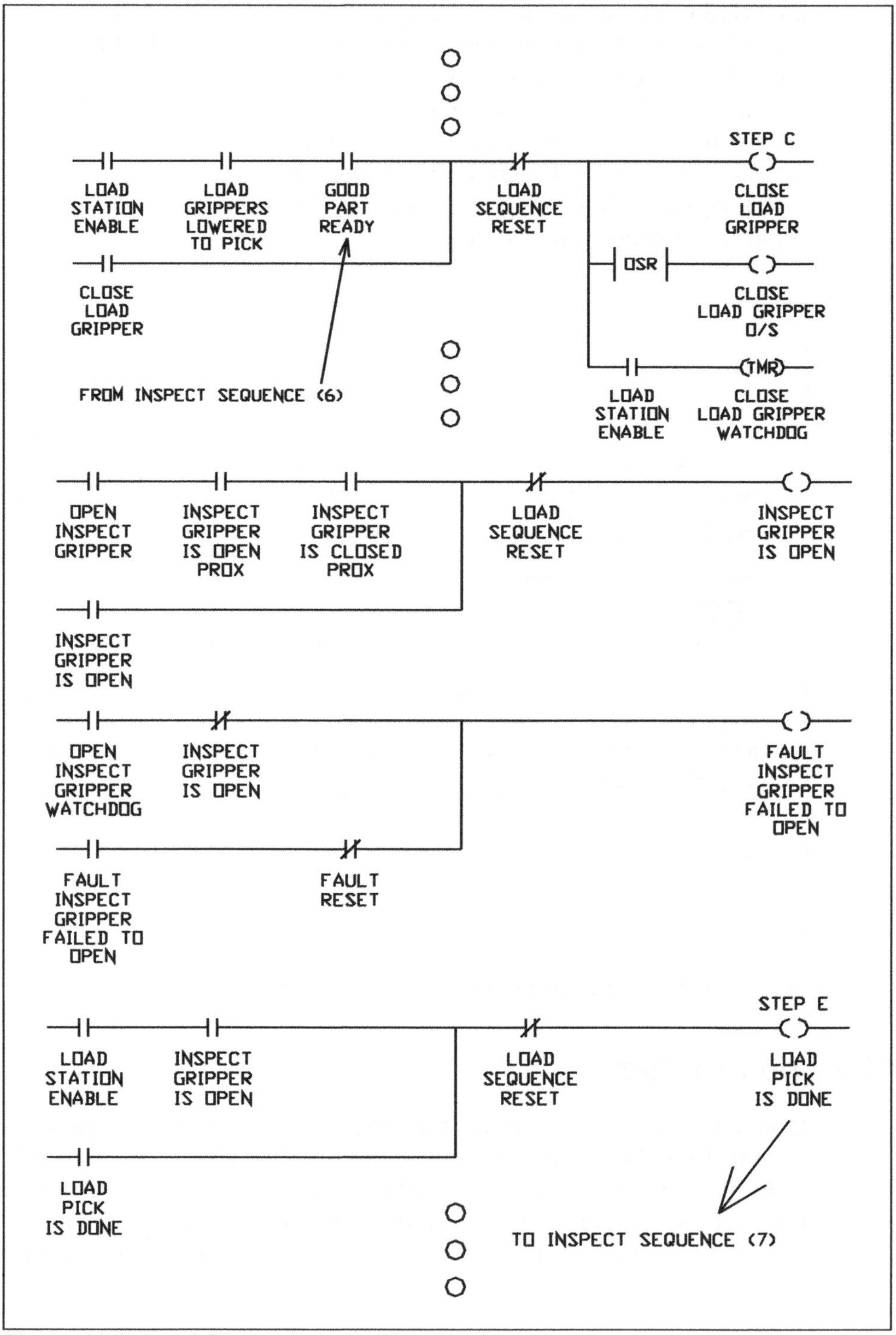

Figure 11-3. Load Station Interlocking Logic

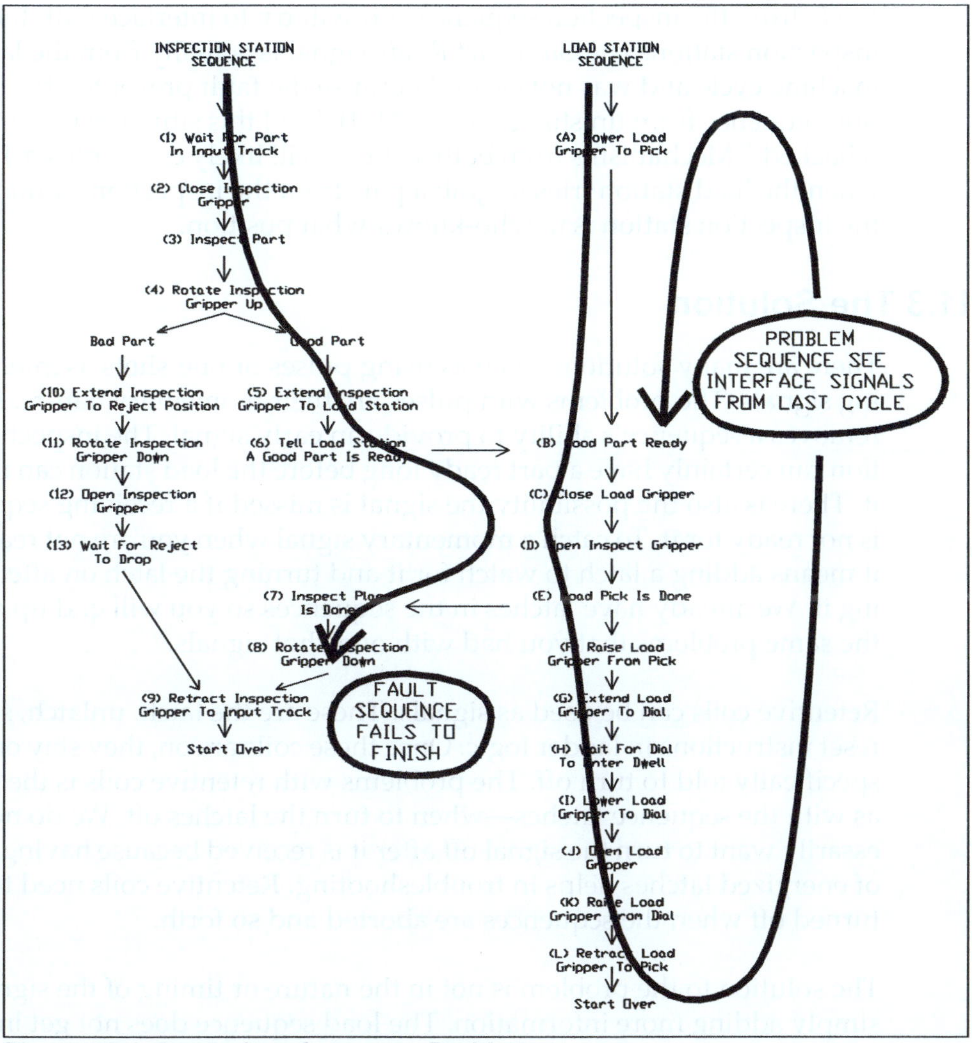

Figure 11-4. Dependent Sequence Problem

step 8 will remain on. The motion complete rung of step 8 is off, which is what triggered the fault. The load station sequence will continue and finish, since the fault occurred after the part transfer and didn't affect the load station's operation. The fault will also drop the *Run System* latch, which will normally prevent the load station from restarting. It is possible the load station will have a chance to restart before the run system latch can stop it. The chance is given when the remaining load sequence after the part transfer is relatively short compared with the remaining inspection sequence.

The problem occurs if the load station sequence restarts while the inspection station is faulted or more likely going to fault (watch dog timer has not yet expired). The load sequence will still see the interlock *Good Part*

Ready from the inspection sequence and will try to interface with the inspection station. The *Good Part Ready* signal is actually from the last machine cycle and was not cleared because the fault prevented the inspection sequence from finishing (Figure 11-4). I call this situation being "bushwhacked." Mechanisms from both stations will likely crash into each other when the load station tries to grab a part from the inspection station, and the inspection station is in who-knows-what position.

11.3 The Solution

There are many solutions, such as using pulses or one-shots as interlocking signals. The problems with pulses are their momentary nature. It removes a sequence's ability to provide an early signal. The inspection station can certainly have a part ready long before the load station can receive it. There is also the possibility the signal is missed if a receiving sequence is not ready for it. To catch a momentary signal when you are not ready for it means adding a latch to watch for it and turning the latch on after finding it. We already have latches in the sequences so you will end up with the same problems that you had with one-shot signals.

Retentive coils can be used as signals. These are the latch, unlatch, set, or reset instructions in ladder logic. Once these coils are on, they stay on until specifically told to turn off. The problems with retentive coils is the same as with the sequence latches—when to turn the latches off. We do not necessarily want to turn the signal off after it is received because having a trail of energized latches helps in troubleshooting. Retentive coils need to be turned off when the sequences are aborted and so forth.

The solution to the problem is not in the nature or timing of the signal but simply adding more information. The load sequence does not get into trouble until it resets, starts over, and ends up back at the interface to the inspection sequence (step B in Figure 11-1). At this point we must decide if the signal *Good Part Ready* is valid or not. The signal is only misleading if the sequence latches below it are on, indicating the sequence is past the interface point. When the downstream latches are off, we know the sequence is at the interface point and not beyond it. By adding downstream information, we can make the *Good Part Ready* signal accurate. The logic correction is shown in Figure 11-5; compare this rung with the first rung in Figure 11-3.

11.4 Other Problems

What if the inspection sequence faults before the part-transfer interface? In this case, the load sequence simply waits for the *Good Part Ready* signal.

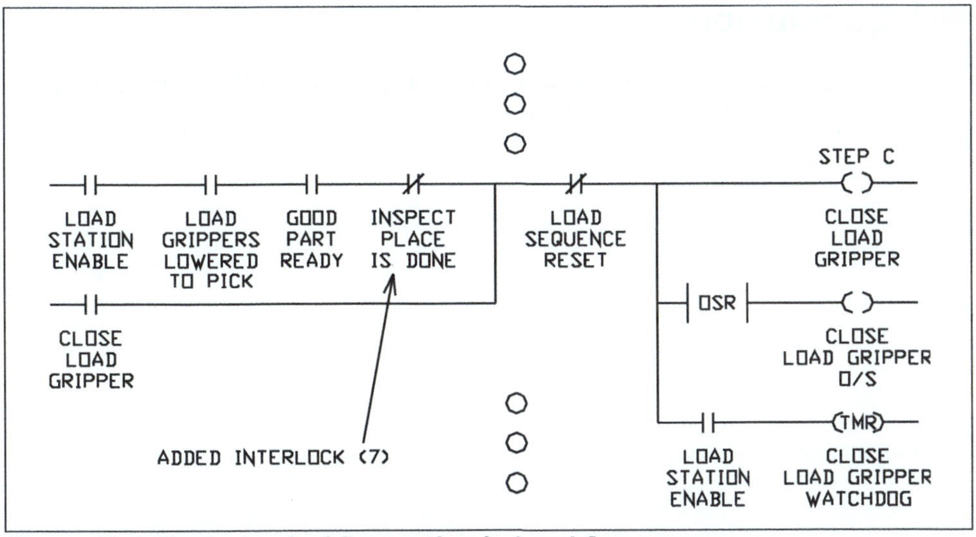

Figure 11-5. Bushwhacked Prevention in Load Sequence

What if the load sequence faults before the part-transfer interface? In this case, the inspection sequence waits for the *Load Pick Is Done* signal.

What if the load sequence faults after the part-transfer interface? It looks like the inspection sequence can get bushwhacked, so additional logic is needed as shown in Figure 11-6. The old logic is shown in the last rung of Figure 11-2

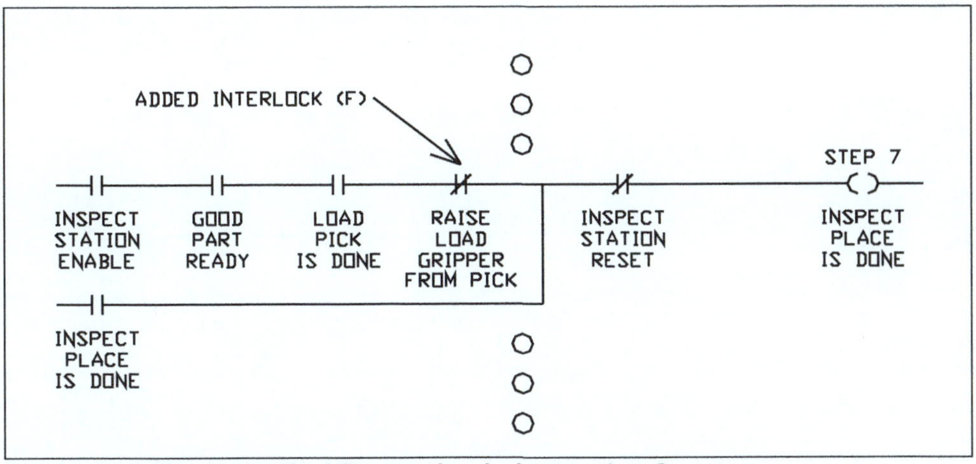

Figure 11-6. Bushwhacked Prevention in Inspection Sequence

The final question is, what happens if a fault occurs during the part-transfer sequence? The load sequence will fault since the mechanisms are triggered and monitored from this sequence, and the inspection sequence will simply wait for the *Load Pick Is Done* signal.

11.5 Conclusion

Always ask yourself the five what-if questions when interlocking sequences. Not all interlocking is as easy as the example illustrated here.

12 Machine States

12.0 Introduction

The introduction to this book talked about the issues that the machine control logic has to address. The operator interface, in particular, is rather important. The operators and maintenance personnel need to perform their jobs as effectively as possible to keep the equipment producing. The machine control logic can greatly help out by making the operator interface easy to use, clear to understand, informative, and as intuitive as possible. The operator interface also needs to include features to assist with setup, calibration, debug, and recovery from problems. Many operator interface functions are a given, such as starting and stopping, but others may not be apparent until you actual see the equipment in operation and experience the problems yourself. To help understand the requirements of the operator interface, we can look at the equipment from the perspective of the operators and maintenance personnel. From this view, we can ask ourselves, "How do I restart the machine after the product feeds itself wrong?" and so forth, developing a list of items to address.

Other issues require that the machine control logic be easy to create to save development time, and be comprehensive to save commissioning time by avoiding rewriting and adding logic. Anticipating and handling every conceivable problem defeats these requirements because it is too much to think about, and even with many years of experience, you probably will not be able to anticipate every problem. The best method is to provide generic approaches with simple options through the operator interface. These generic approaches are called *fault recovery tools*. They allow the operators and maintenance personnel to handle virtually all the possible problems. Logic can always be added to handle special cases. These are usually found while starting up and commissioning the equipment. Special logic is rarely required because the methodology of Cascading Logic provides good fault recovery. Cascading Logic also lends itself to adding special recovery logic because of how the logic is encapsulated and organized. Fault recovery tools include alarm annunciation, manual mode operations, and individual control of each station on the machine.

12.1 Machine States

Operator interface features and fault recovery tools were discussed in the preceding chapters. However, to put the operator interface into perspective, we need to look at the equipment from the view of the operators and maintenance personnel. This time, let's look at all the different conditions we can find the equipment in, such as running production, sitting idle, or stopped due to a problem. Let's define a machine state as one of the possible machine conditions. We can use machine states to anticipate the conditions the machinery can end up in and to consider what an operator has to do to get the machine running again. Categorizing the possible machine conditions into machine states helps us anticipate problems and plan operator interface tools.

It is also important to understand the possible machine states to provide routes between them. The machine's whole purpose is to run production, and the operators need methods and tools to help accomplish this goal. Downtime is reduced when the machine is easy to operate and placed into production quickly, no matter what its current condition is. The theory of machine states thus has two parts. The first is to categorize all the possible operation conditions into a few definable states. The second is to define and make routes available between these states.

Operators use machine states like a road map. The operator first determines the current machine condition, then using one of the provided routes, navigates the machine back into production. The machine needs to provide feedback to assist the operator. Simple machines can use pilot lights and larger machines can have text displays on human–machine interfaces (HMI).

Before we categorize the possible machine conditions into machine states, let us further define what a machine state is.

12.2 Machine State Concepts

- They have to be few in numbers. We want to save time and make it easy for the operator. With too many states, you will be programming until midnight and the operator will be spending more time reading manuals than filling part hoppers.

- They have to be definable in terms the machine can understand through available sensors or derived through logic. You are not part of the machine, so the machine itself needs to determine what state it is in.
- They have to be useful. The whole theory of machine states is to get the machine running from any possible condition. This is to minimize downtime and bring fame and glory to you as the programmer.
- They have to be stable. Unstable states may be dangerous by causing unintentional machine motion or by misinforming the operator.
- The machine can be in only one state at a time. Overlapping states may be dangerous by causing confusion and unwanted machine motion.
- There needs to be at least one route out of each state. Turning the machine's disconnect off to reset the machine control logic is not an acceptable route. Using a support computer to look at and correct the logic is only acceptable during equipment startup and commissioning.

Figure 12-1 shows seven useful machine states and the routes between them.

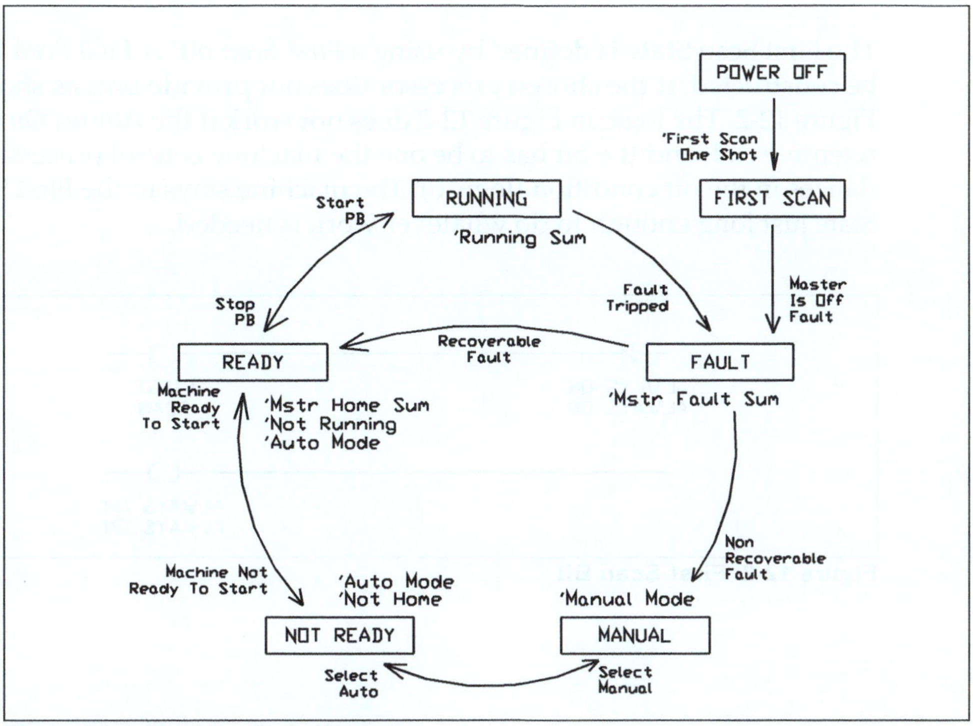

Figure 12-1. Machine State Diagram

12.3 Power Off State

This state is insignificant. The machine cannot solve any logic with the power off, so why write any logic for it? This state is included for completeness, and more importantly, as a reminder that sooner or later power will be applied and hence the next state.

12.4 First Scan State

In this state, power to the machine just came on and the machine controller starts solving its logic. This state is useful for initializing the machine. Machine sensors are usually operational at this point, providing the machine with eyes and ears. Motive power is not on. Motive power is applied only through the master control circuit. The only way to energize the master control circuit, and thus allow motors to rotate and solenoid valves to fire, is for someone, not the machine, to turn it on.

The machine should not receive motive power until it is ready to handle it. This can be guaranteed by adding an output from the machine controller through a relay to the master control circuit. This relay can de-energize the master control circuit and keep it de-energized until the machine is ready for motive power. In many instances, there is not much initialization to do. In this case, the fastest time anybody can turn on the main disconnect switch and then run over and press the *Master Start* button is sufficient.

The First Scan State is defined by using a *First Scan* bit. A *First Scan* bit can be constructed, if the chosen processor does not provide one, as shown in Figure 12-2. The logic in Figure 12-2 does not work if the *Always On* bit is a retentive coil, and the bit has to be one the machine control processor initializes in the off condition (logic 0). The machine stays in the First Scan State just long enough to do whatever work is needed.

Figure 12-2. First Scan Bit

Here is a list of some tasks accomplished by the First Scan State.

- Download configuration data to special I/O modules.
- Synchronize values between the programmable logic controller and the operator interface panel (HMI). The operator interface panel, or the programmable logic controller, is capable of holding the current machine parameters, such as temperature set points. You can usually select if the HMI will upload or download its values from or to the programmable logic controller on startup.
- Set mode selection logic (manual, step, single-cycle, and automatic) to the default manual mode.
- Set/reset shift registers, data pointers, and other memory logic.
- Establish communications with other smart devices or area controllers.
- Trigger solenoid valves to align the spools to match the current position of the corresponding cylinders. Generally, you do not do much with valves here because the machine should not move, just because the master control circuit is energized. Motion should occur only through deliberate actions of the operator, such as pressing motion-initiation buttons in manual mode or pressing the *Cycle Start* push button in automatic mode.
- Briefly display company logos.

12.5 Fault State

The machine is in the Fault state whenever a fault is active. The only exception is when the machine is briefly in the First Scan state. Faults are problem conditions detected by the machine. They usually stop and/or prevent the machine from running when active, and they should always be announced to the operator. Faults can happen at any time, so the Fault state can be entered from any other state. An example of this is an emergency stop button. It can be pressed at any time to de-energize the master control circuit and generate a *Master Is Off* fault.

The Fault state is the next state entered after the First Scan state. This is because I define a de-energized master control circuit as a fault condition, as shown in Figure 12-3. When incoming power to the machine is applied (main disconnect switch is turned on), the master control circuit is de-energized. The master control circuit is fail-safe to de-energize with loss of power and can only energize when specifically told to do so, that is, a deliberate operator action. Exceptions to this occur when a programmer changes the operating mode of the machine controller with a support computer, or somebody uses the mode key switch on the controller. These cases are not associated with the operator but with people who should know the consequences and who should warn people around the equipment.

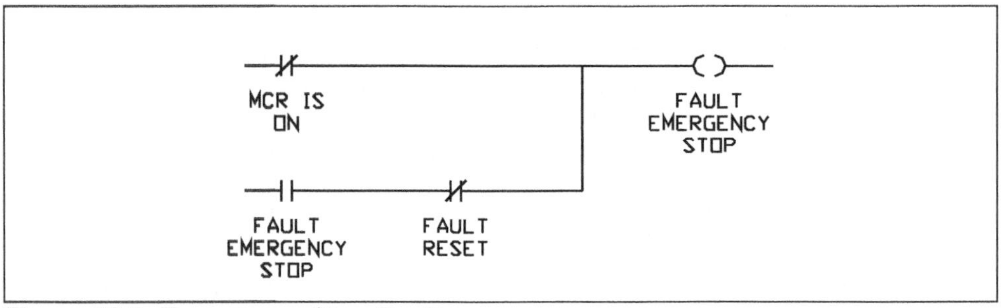

Figure 12-3. Emergency Stop Fault

The logic shown in Figure 12-4 is an example of a Sequence fault. It is used to watch a motion and energize if the motion fails to complete within a time limit ("dog" is short for "watch dog timer").

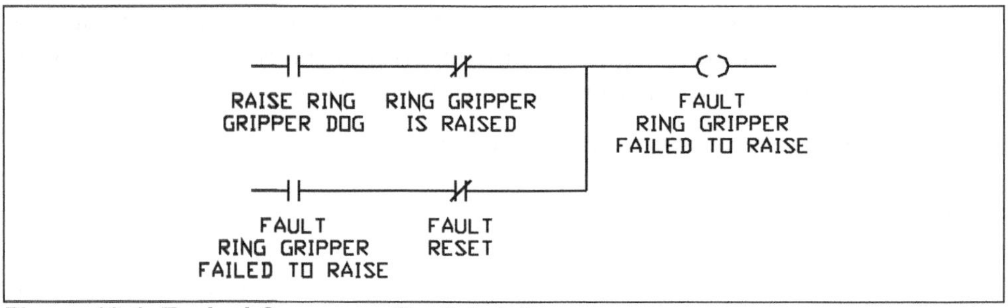

Figure 12-4. Typical Sequence Fault Latch

All the individual fault latches can be summed up into one bit called the *System Fault Summation*. The machine is in the Fault state whenever the *System Fault Summation* is on. With this bit you can do the following:

- Freeze the sequence logic to effectively stop the machine in its tracks. This allows the operator to examine the machine in the faulted position to see what went wrong. I have seen many machines return to their home position after a fault occurs, leaving everybody guessing what went wrong. Frozen logic keeps the machine from further damaging itself and prevents the machine from skipping to the next step when sensors are accidentally triggered. Freezing the logic has two other benefits. First, the machine now has the capability to resume from where it stopped, reducing downtime. I call this a recoverable fault, as illustrated in Figure 12-1 as the path from the Fault state to the Ready state. An example of a recoverable fault is where the operator enters the machine and repositions the product, closes the guards, and then restarts the machine. Second, the intimate details of the logic are locked in time. The logic can be viewed to see what went wrong. Examining frozen logic is a big help during the startup commissioning phase and when providing modem support.

- On machines with multiple stations, the faulted station can be frozen, and the remaining stations can be cycle stopped. This allows the non-faulted stations to enter stable conditions and allows quicker restarts. Having a pick/place station finish placing a part is always better than leaving the part hanging from vacuum cups, especially if the operator needs to enter the machine. When an operator enters the machine, the motive power, including the main air supply, is dumped for operator safety. Dumping air pressure can, by itself, cause additional problems.
- Helps direct appropriate messages to the operator. The message control logic can be freed from scanning all the messages to scanning only the fault messages. This will provide timely information to the operator by displaying only relevant information. In theory, only the root cause of the problem should be directed to the operator. In many cases a bad machine condition can have a cascading effect, tripping multiple faults. Multiple faults and unrelated information can be controlled through the use of logic freezing and message control.
- Prevent the machine from starting or resuming until the operator handles the faulting condition. After a fault stops the machine, the operator should inspect the faulted area and then determine the best course of action. The fault message can be backed up by the equipment's manual. The manual can include a description of the fault, most likely causes, including the sensors involved, and remedies to restart the machine. I always include a message number in the fault message for quicker reference to the equipment manual. I find it is easier for an operator to remember a number rather than a lengthy message. Having an alarm history screen to display the last faults also helps.
- Prevent initiation of motion in manual mode until the fault is cleared. This is an attempt to force the operator to examine the faulted area before proceeding. It can also prevent the more intelligent motions, such as servos, from responding to operator request if they have a problem.

12.6 Manual Mode State

The machine is in the Manual Mode state when the operator selects manual mode and no faults exist. A fault can be generated when the machine is running and manual mode is selected. The fault can either freeze the machine operation or cycle stop the machine. This is why no path is shown in Figure 12-1 from the Running state to the Manual Mode state because it is routed through the Fault state.

Manual mode is useful by providing the operator direct access to machine motions for testing and machine setup, and moving the machine to its starting (home) position. Interlocks are incorporated to protect the machine from inadvertent operator action. For instance, a pick/place mechanism can be prevented from moving horizontally if the grippers are not raised.

Manual mode offers enormous recovery options. The operator can choose to enter manual mode after the machine stops because of a fault. The operator should always inspect the faulted area before deciding what to do. In some cases, the operator may not be able to resume machine operations; these cases are called nonrecoverable faults. An example of a nonrecoverable fault is when the product needs to be removed from the faulted station. The faulted station sequence can be aborted and the machine placed in manual mode. The operator can then place the station back to its starting (home) position and then restart the machine to allow the station to repeat. Take a station that places one component part into a main assembly. The once faulted station simply starts over and places a new part into the main assembly. This results in saving the main assembly. The whole theory of fault recovery is providing the operator and the machine every possible means to save the product and/or keep the machine running.

Manual mode also prevents the operator from starting automatic operations in contrast with automatic mode, which prevents the operator from initiated motions while the machine is running, a scary thought.

12.7 Ready and Not Ready States

The Not Ready state is the direct opposite of the Ready state. In a nutshell, the Not Ready state is used to tell the operator any reason that will prevent the machine from starting. The Ready state means all conditions are good, so if the operator presses the *Cycle Start* push button, the machine will start automatic operations. When the machine is in the Ready state, a single message can be displayed to the operator, "Ready To Start," to indicate all is well.

The Not Ready state is used to direct the operator in preparing the machine to start automatic operations. This state can direct messages to the operator to indicate which motions need to be repositioned to their starting points (not home messages). Since the machine has to be in manual mode to allow the operator to reposition motions, these message can also be displayed in manual mode, provided no faults exist. This state can also tell the operator to start auxiliary equipment, such as hydraulic pumps, scrap vacuum units, and dust collectors.

For the machine to run automatic operations, the following conditions need to be true. The machine can also determine which state it is in by examining these conditions.

The machine needs to be in its starting (home) position. The position of each motion can be checked and summed up into one bit called *System Ready.* When the machine is at its starting position, this bit is on. Consideration has to be given to stations on the machine that are frozen, for these stations will resume from where they were stopped and will not necessarily start from their home position.

- The machine needs to be in automatic mode. We do not want the operator moving cylinders around when the machine is commanding them to move. The operator has manual mode to initiate motions.
- No faults exist. All the faults need to be dealt with before the machine can be started.
- All auxiliary equipment the operator needs to start has to be on before the machine can be started.

12.8 Running State

The Running state is shown at the top of the circle in Figure 12-1 because this is the most desired state. The machine can enter the Running state only from the Ready state when the operator presses the *Cycle Start* push button.

The Running state is determined by examining the control latches of each station. The control latches indicates if the station is running, frozen, or idle. The control latches are discussed in Chapter 6 on controlling sequences. All the control latches can be summed up into a *Machine Running* summation bit and be used to inform the operator that the machine is, or will be, moving. Will be moving is an important safety consideration. A machine may appear to be idle, but then move suddenly—a good reason for guard doors. Sudden movements can be caused by the machine waiting for a part from a feeder bowl, then firing cylinders to grab the part when it shows up. Another case is when a cylinder jams but before a watch dog timer can trigger a fault.

The running conditions are used to operate in-cycle lights and to display a "Machine Running" message. The in-cycle lights can be made to blink if a station has a fault, while the remaining stations are cycle stopping. Once the in-cycle lights are off, the operator can enter the machine with out disrupting the remaining stations. If guard door locks are used, they can be unlocked after all the stations have stopped.

12.9 Other Possible States

Two other common conditions that equipment experience require careful analysis to deal with effectively. These conditions occur when motive power is removed and restored. Motive power includes electrical power to solenoid valves and motors, pneumatic and hydraulic pressure to cylinders, and so forth. These conditions are detected by monitoring the master control circuit and pressure switches.

These two conditions are not shown in the diagram in Figure 12-1 because I consider them special conditions of the Fault state. They can be thought of as separate states, if this helps you remember these conditions.

Some things to do when losing or restoring motive power:

- Freeze sequences. This allows a chance at resuming operations once power is restored and preventing watch dog faults from triggering.
- Servo operated motions can be commanded to repeat their last move after the machine is restarted. This is especially easy if the servo moves are absolute positions.
- De-energize certain solenoid valve outputs to prevent motion after power is restored. The valves themselves will de-energize when power is removed, but what do you want the valve, thus the motion, to do when power is reapplied? A careful analysis of the motion, valve type, related motions, and gravity are used to answer this.

12.10 Conclusions

The biggest emphasis of machine states is organizing and presenting messages to the operator. Some day, machines will be equipped with enough sensors and brain power to allow George Jetson to run them (he shows up only long enough to press the start push button). Until then, operators are an integral part of machine operations. The machine needs to keep them informed and provide them will sufficient tools and options. The operator can then complete the control loop by deciding what to do and initiating appropriate action.

13
Message Logic

13.0 Introduction

In a perfect world, machinery stops only when told to do so. In the real world, mechanisms bind up and product sticks. Let us define *downtime* as all the time the machinery is idle when it should be producing. Our goal is then to minimize downtime. This can be accomplished through a wide variety of methods as listed below.

- Robust and reliable mechanisms
- Robust and reliable programs
- Product conforming to specifications
- Operator and mechanic training
- Knowing what went wrong and where problem occurred

No one method by itself can completely eliminate downtime, so a combination of methods is required. However, problems still occur, even with all bases covered. A simple way to keep the operator informed provides the most bang for the buck keeping in mind the practicality and cost of each method. It involves only relatively inexpensive hardware for annunciation and a little extra programming.

All previous chapters emphasized the need to keep the operator informed. This chapter will show how to construct the message logic.

13.1 Motion Object

A motion object is used to encapsulate each step a machine is required to perform. A basic principle of the motion object is that for every initiated action there are two possible outcomes. One outcome is the successful completion of the action, and the other is the failure of the action. Figure 13-1 illustrates a typical motion object. Chapter 2 on sequences and motion objects explains the motion object in more detail. Its possible to alert the operator to the exact nature of a problem in a timely manner when each machine action is monitored. Machine actions are not limited to extending cylinders or starting motors but can also include such actions as sending data to other devices.

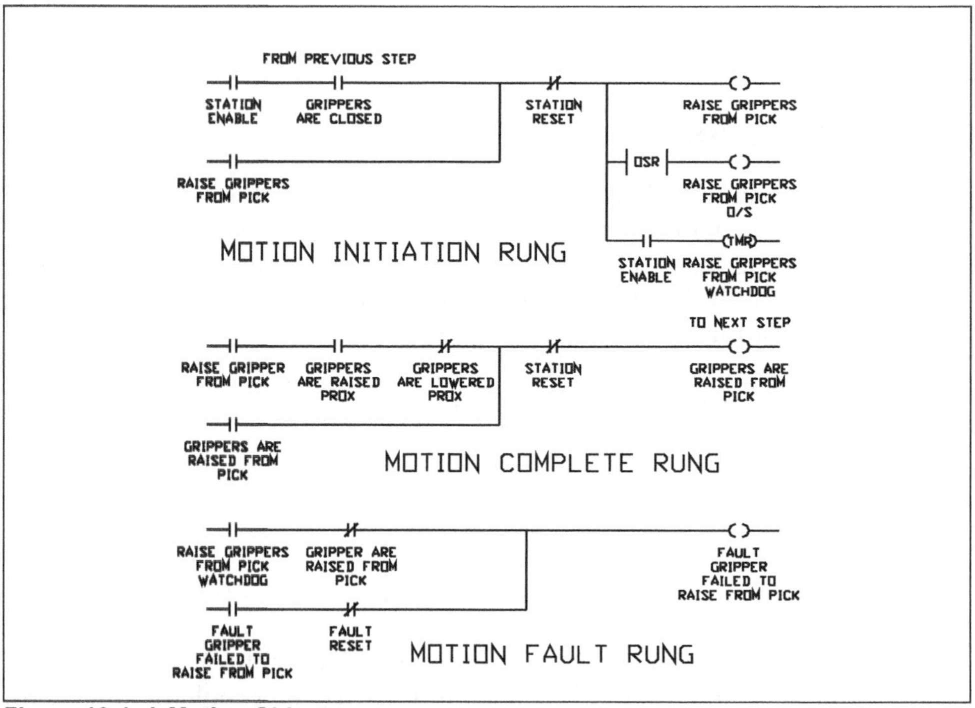

Figure 13-1. A Motion Object

The fault latch of the motion object is used to capture the failure of the initiated action and forms the basis of message logic. It is the fault latch that knows there is a problem, and it is the memory element that is used to trigger a message. All that is needed is logic to scan the fault latches and display a message for the active faults.

13.2 Message Displays

Message displays can be discrete devices or an HMI object. They have the functionality to display a message corresponding to whatever number you send it. If you change the number, a different message appears. The discrete devices typically receive the message number via digital input/output lines. Eight input/output lines, each interpreted as a binary number, can trigger up to 255 messages. Message displays on HMIs receive their message number over a data link. An integer value (16 bits) can select up to 65,535 messages.

13.3 Message Display Logic

A message bit is any coil that is required to display a message whenever it is active (on). Typical message bits include fault latches from the motion objects and other coils used to indicate machine status, such as machine running or low product hoppers. The machine program can then scan all the message bits and send a corresponding message number to the display unit when an active message bit is found. After giving the operator time to read the message, typically three seconds, the scanning of the message bits continues. If another active message bit is found, then a new message is displayed. Once the message-bit scanner reaches the end of the message list, it simply starts over at the beginning. If the same message bit is still active, its corresponding message is again displayed. A message display programmed in this manner is capable of displaying messages for many concurrent faults.

The key to message logic is to group the message bits together, leaving room for growth. Grouping the message bits together allows them to be scanned. There are several methods available to scan them. One way is to use an instruction that provides a pointer or index number to the next bit within the group that is on. The instruction output will yield 20 if the 20th bit is on. The output of the instruction can then be fed directly to the message display. Another method uses indirect addressing, where the number contained by the register in the logic refers to another register, where the data is taken from. An example of indirect logic is shown in Figure 13-2.

The rungs in Figure 13-3 uses indirect addressing to scan the message bits. The message number (N7:0) selects which message bit to look at. If the selected message bit is on, the *Message Found* bit (B13/0) is latched, and the message duration timer starts. The latch stays on until the message duration timer expires. The message number is passed to the message display (N7:1) when the *Message Found* latch first turns on. This will prevent the message display from tracking the message number when the message number rapidly advances to the next active message bit.

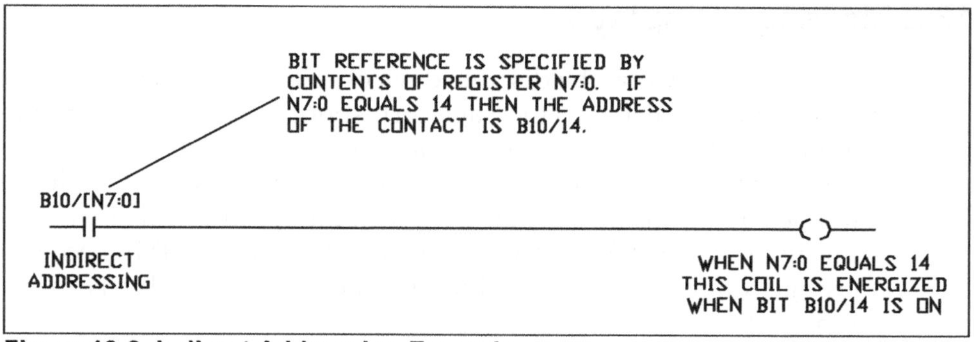

Figure 13-2. Indirect Addressing Example

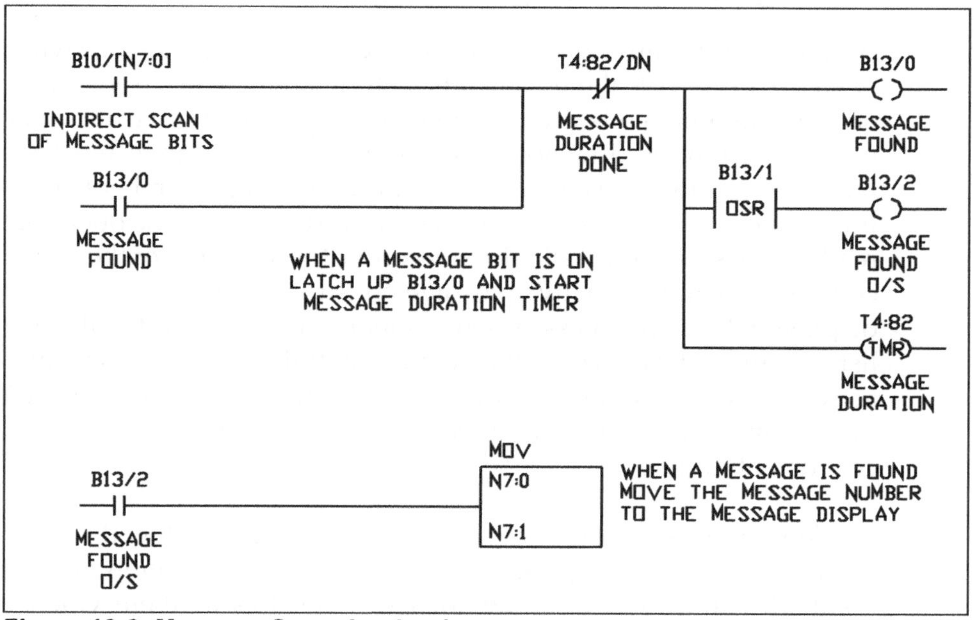

Figure 13-3. Message Scanning Logic

The message-number control is shown in Figure 13-4. The message number is incremented by one, to point to the next message bit when no active message is found. The message duration timer is used to keep the *Message Found* latch on, to prevent the message number from being advanced. This will give the operator time to read the message. The message number is reset back to the first message bit after the last message bit is scanned. In this example the first message bit is 0 and the last is 214.

It takes one program scan for each increment of the message number. This is a problem only when scanning large quantities of message bits. Let's say the program scan time is 20 milliseconds and we have 214 messages. It will require 4.28 seconds to scan all the message bits. Instructions that can scan a group of bits do not have this problem because they provide a

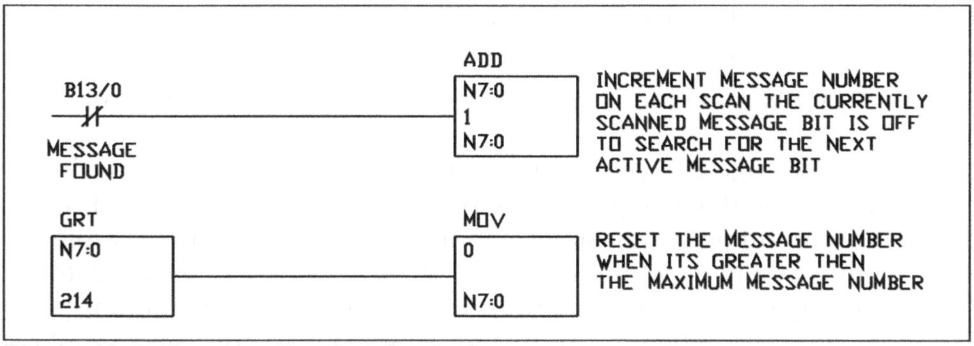

Figure 13-4. Message-Number Control

pointer to the next active bit within one scan of the program. When using indirect addressing, we can arrange the messages into subgroups, which follow the possible machine states. Figure 13-5 shows how the *System Fault Summation* can be used to restrict the message number to scan only the fault latches. In this example, the fault latches are arranged between 106 and 214, thus cutting the message scan time approximately in half. Other machine-state bits, such as *Machine Running* and *Manual Mode,* can also be used to redefine the message-bit scan.

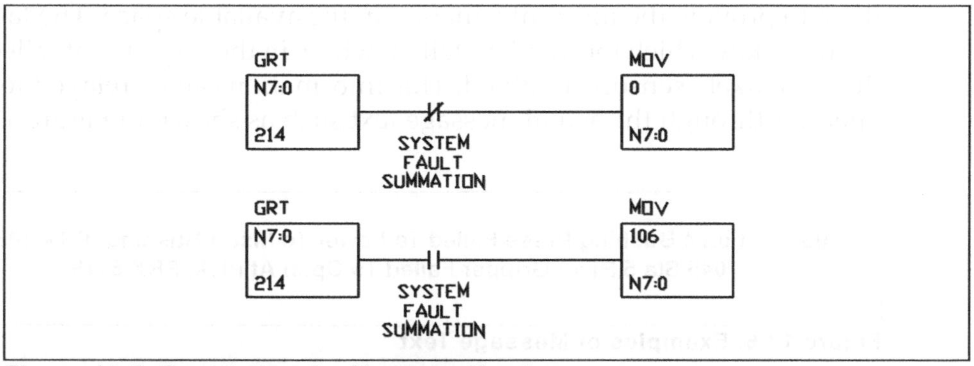

Figure 13-5. Restricting the Message Number

13.4 Types of Messages

Downtime reduction through keeping the operator informed boils down to displaying a message for anything that prevents the machine from starting and anything that stops the machine. The different type of messages can be grouped as follows.

- Fault Messages (fault latches)
- Status Messages (running, manual mode, not ready)

- Not Home or Not Ready Messages (grippers not open, pusher not retracted)
- Warning Messages (low product hopper, exit conveyor full)

The message groupings follow directly from the possible machine states, as discussed in Chapter 12 on machine states. The concept of discerning the current operation status of a machine into only one possible state is important in deciding which messages to display to the operator. Messages to alert the operator to mechanism not in their home position are meaningless when the machine is running. The not home or not ready messages can be conditioned by the *Machine Running* summation, preventing their appearing when they are not needed. In a similar fashion, the *System Fault Summation* can be used to block all nonfault messages. It does not matter if a feeder bowl is low when a pick/place mechanism failed to extend.

13.5 Message Text

The message display quickly becomes the focal point on the machine. New operators especially depend on it to guide them in operating unfamiliar equipment. It is not enough just to have messages, but to phrase them to provide the most information in the available space. The fault latches know which mechanism failed, where in the sequence it failed, and the most likely sensors involved. This information can be relayed to the operator through the actual message text such as shown in Figure 13-6.

037 Station 4 Bushing Press Failed To Lower To Insert Bushing, PRX 10345
045 Sta 5 Rivet Gripper Failed To Open At Pick, PRX 5118

Figure 13-6. Examples of Message Text

The key points to a good message:

- Identify the machine section (Station 5)
- Identify the mechanism (rivet gripper)
- Identify what the mechanism failed to do (failed to open)
- Identify the point in the sequence (at pick)
- Identify the sensor or sensors (proximity switch 5118)
- Provide a message number (045)

The message number has several uses. The first is to identify the internal coil within the machine program that is responsible for triggering the message. The second is to provide a reference number.

Organization of the message logic and message bits makes identifying the internal coil easy. The message scanning logic provides the message display with the number 20 if the 20th message bit is on. The 20th message bit should then have an address such as B13/20 or XXX020, where XXX does not matter. Any offset between the displayed message number and the actual internal coil can be explained in the equipment manual. Providing a reference to the internal coil helps in debugging the message logic and the machine for the programmer and maintenance personnel.

Program all unused or reserved messages with a simple message that includes the message number, such as "message 935". Rather than having to guess at the cause for a blank display, you will have a pointer to the problem.

I find it easier to remember a number, rather than a whole line of text. An operator may remember that the gripper failed at the load station, but may not remember if it failed while picking or placing the part. Having a number for quick reference solves this problem. I can hear the operator saying, "The swing arm stopped and message 36 came up." The message number also makes it easier to find additional information in the equipment manual. The manual can include recovery procedures and likely causes of the problem.

To make writing the troubleshooting section of the equipment manual easier, keep a printout of the messages with you when commissioning and debugging the equipment. The printout provides a vehicle for capturing information when faults occur. The cause of a problem, what to look for, how to access, and how to recover can all be noted while the information is fresh in your mind.

13.6 Message Logic Payback

Many programmers add message logic as an afterthought or to fulfill a requirement. I find messages such a useful debug tool that I make message logic the first part of the machine to work. Incorporating the fault latch within a motion object makes programming messages easy and straightforward with only a little extra effort. The payback is immediate and often. When commissioning and debugging equipment, the fault messages directs you into the sequence where the problem occurred. Debugging is further enhanced by the cascading effect of the sequence latches. You can follow the trail of energized latches and see why a particular latch failed to energize.

The benefits of using message logic as soon as possible becomes obvious when considering the costs involved. Programming extra fault latches in a motion object is a relatively small effort, especially when you are geared up to do so and are programming in the comfort of your office. The message scanning logic itself is only a handful of rungs. Problems out on the equipment are more difficult and costly to solve. Noise, time pressure, and other distractions slow you down and there are also more people involved. While you are fixing a problem, at least one other person is waiting so now the cost doubles. Messages go a long way in helping to reduce your debug time and the end users' downtime.

Appendix A

A.0 Sensor Response Time

The following figures show how to calculate the response time of a proximity switch in combination with a translational and rotational flag. Its important to know how long a signal is energized and de-energized to determine if the programmable logic controller can see the signal. The scan time of the controller has to be greater than the sum of the sensor response time plus the input module response time.

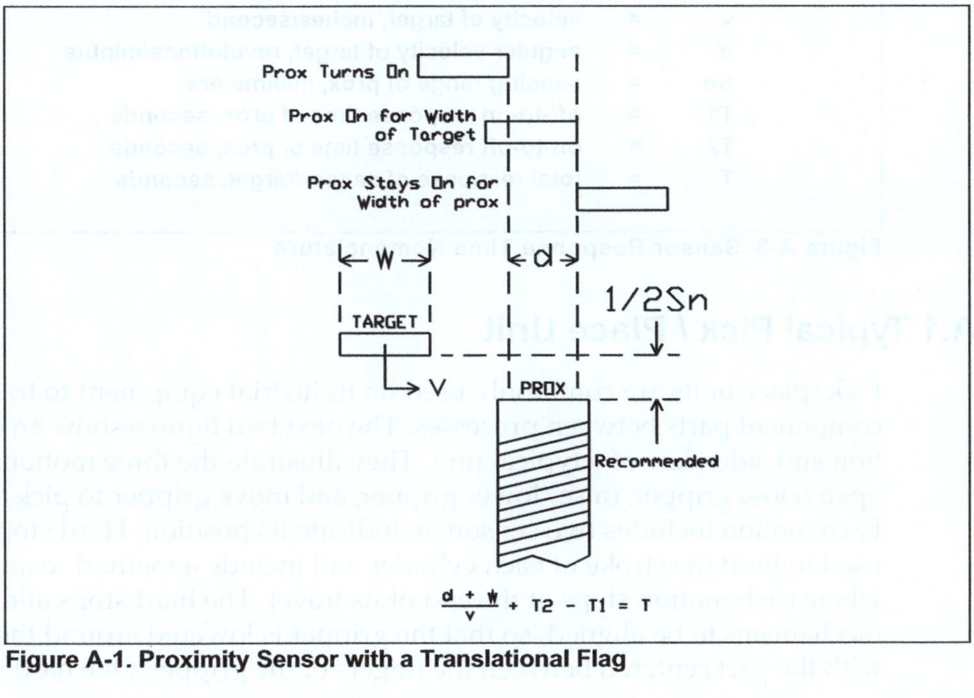

Figure A-1. Proximity Sensor with a Translational Flag

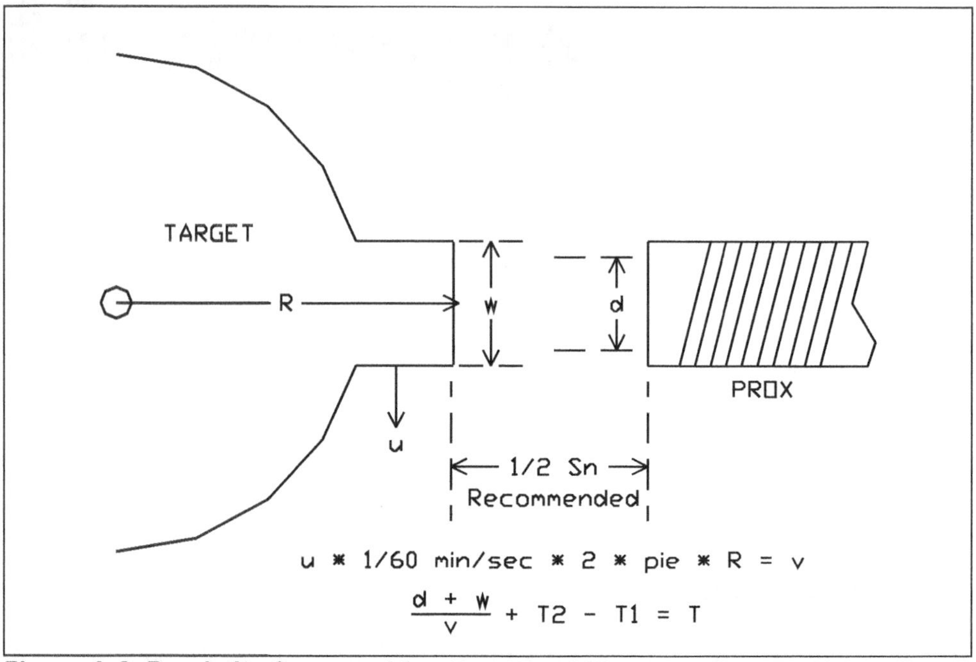

Figure A-2. Proximity Sensor with a Rotational Flag

w	=	width of target, inches
d	=	width of prox (approx. 90% of diameter, consult manufacturer), inches
R	=	radius, inches
v	=	velocity of target, inches/second
u	=	angular velocity of target, revolutions/minute
Sn	=	sensing range of prox, millimeters
T1	=	off-to-on response time of prox, seconds
T2	=	on-to-off response time of prox, seconds
T	=	total response of sensor/target, seconds

Figure A-3. Sensor Response Time Nomenclature

A.1 Typical Pick / Place Unit

Pick/place units are commonly used on industrial equipment to transfer component parts between processes. The next two figures show an elevation and side view of a typical unit. They illustrate the three motions of open/close gripper, raise/lower gripper, and move gripper to pick/place. Each motion includes two sensors to indicate its position. Hard stops are used to limit the stroke of each cylinder and include a method to adjust where each motion stops at the end of its travel. The hard stops allow the mechanisms to be aligned, so that the gripper is lowered around the part with the part centered between the fingers of the gripper. One hard stop is pointed out in Figure A-4, but each motion has two. This example shows a gripper that opens and closes to transfer the part, but many times a vacuum cup is utilized instead.

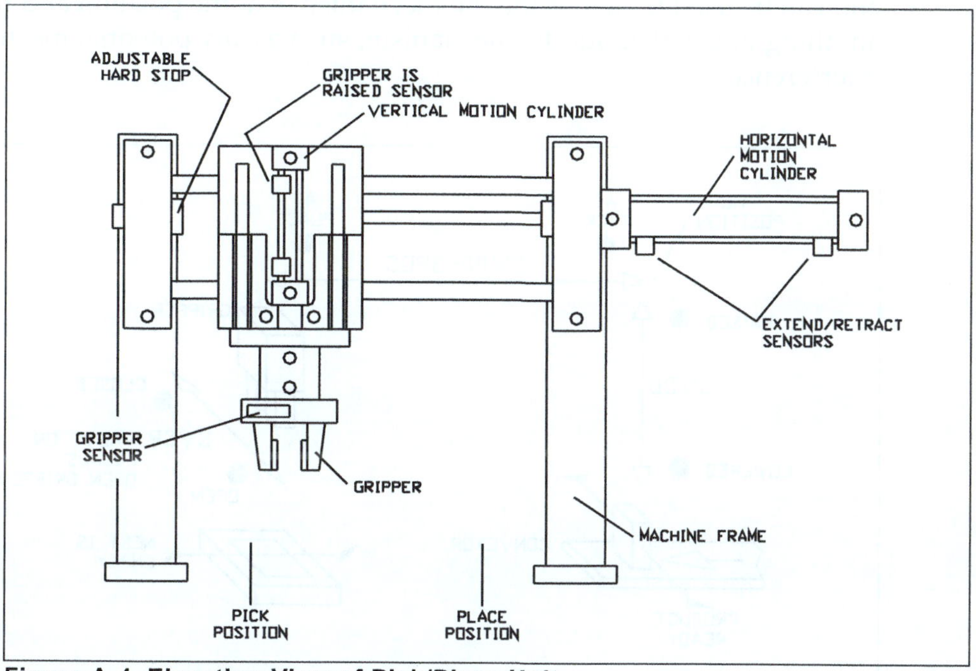

Figure A-4. Elevation View of Pick/Place Unit

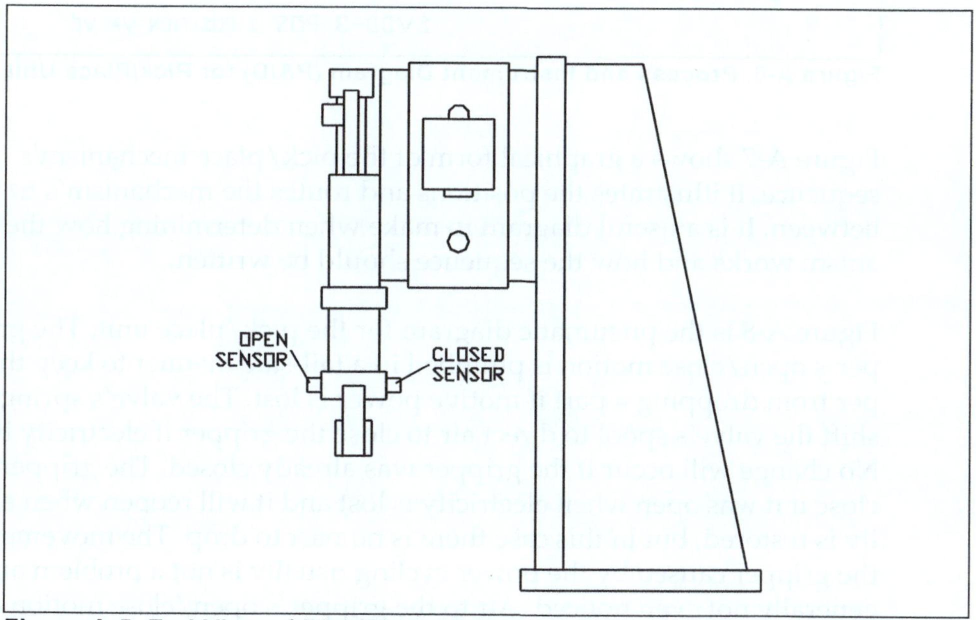

Figure A-5. End View of Pick/Place Unit

Figure A-6 is a process and instrumentation diagram. It shows all the motions, sensors, the type of sensors, and the types of valves used to control the mechanism. It is a useful diagram for programming because it is easy to make, provides a visual reminder of how the mechanism works, and can be used to determine the mechanism's home position and to note anything unusual about the mechanism, such as any potential mechanical interference.

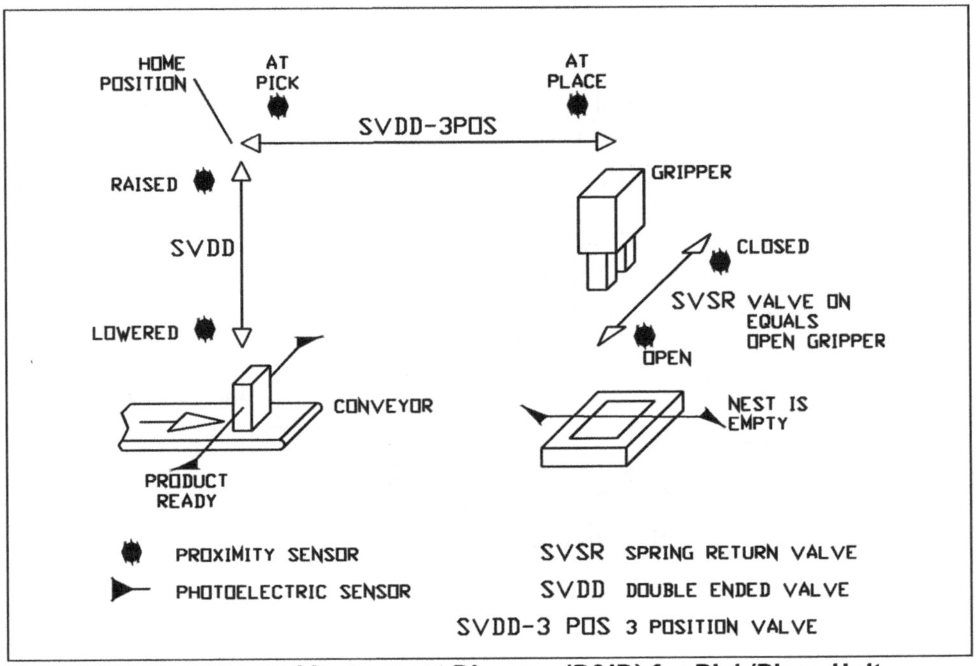

Figure A-6. Process and Instrument Diagram (P&ID) for Pick/Place Unit

Figure A-7 shows a graphical form of the pick/place mechanism's sequence. It illustrates the positions and routes the mechanism's travel between. It is a useful diagram to make when determining how the mechanism works and how the sequence should be written.

Figure A-8 is the pneumatic diagram for the pick/place unit. The gripper's open/close motion is plumbed in a fail-safe manner to keep the gripper from dropping a part if motive power is lost. The valve's spring will shift the valve's spool to direct air to close the gripper if electricity is lost. No change will occur if the gripper was already closed. The gripper will close if it was open when electricity is lost and it will reopen when electricity is restored, but in this case there is no part to drop. The movement of the gripper caused by the power cycling usually is not a problem and is generally not even noticed. Air to the gripper's open/close motion can be plumbed upstream of the machine's air dump valve when the closing of

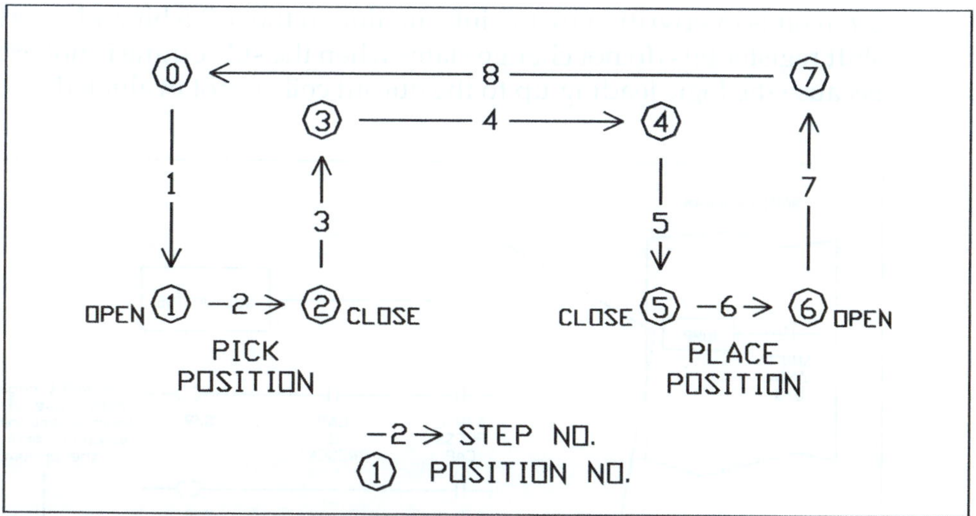

Figure A-7. Pick/Place Sequence

the gripper poses no hazard to a person's fingers, as is usually the case. This will prevent the grippers from dropping parts when air is lost.

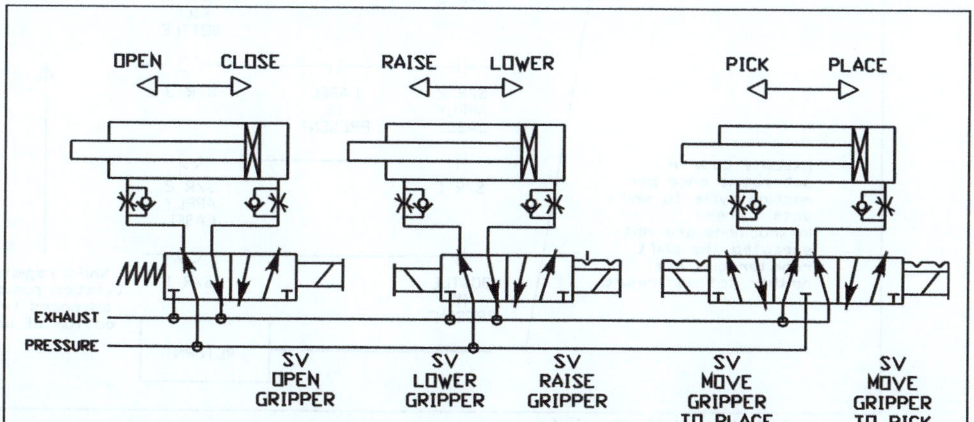

Figure A-8. Pick/Place Unit Pneumatic Diagram

A.2 Shift Registers

The logic in Figure A-9 can be used to advance the status of a single bit as the product advances through the machine. The *Shift Data O/S* bit allows the subroutine to execute just once, when the data needs to advance. The subroutine executes just like standard ladder logic, one rung at a time, starting from the top rung to the bottom rung. The data is advanced from one shift register position to the next because of the scan order of the rungs in the shift register. The top rung, *S/R 7*, is executed first and uses the information from the next rung, *S/R 6*, before the information held by the

S/R 6 bit is overwritten by the information in the *S/R 5* bit and so on. The shift register bits do not change states when the subroutine is not executed because the logic leading up to the output coils is not evaluated.

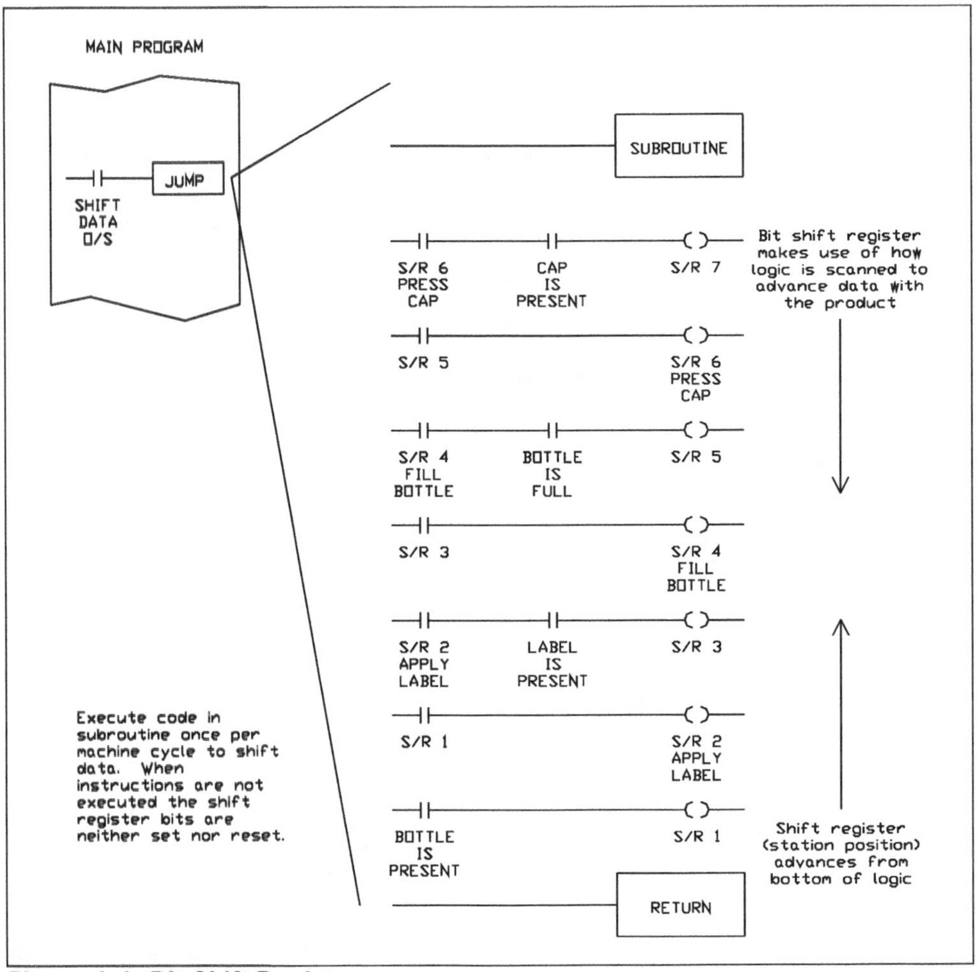

Figure A-9. Bit Shift Register

Words can be shifted instead of individual bits, as shown in Figure A-10. A typical word in a programmable logic control is 16 bits, with the newer controllers providing 32 bit words. A bit can represent each step in the machine's process to track where failures occur, as shown in Figure A-11. The bits of the shift register can be used to trigger counters as the rejected parts are removed from the machine. The counters will tell you the reject rate of each station or process on the machine, pointing out where attention is needed.

Multiple words can be shifted when large amounts of data are tracked with the product, or a pointer can be used. A pointer is a single word that

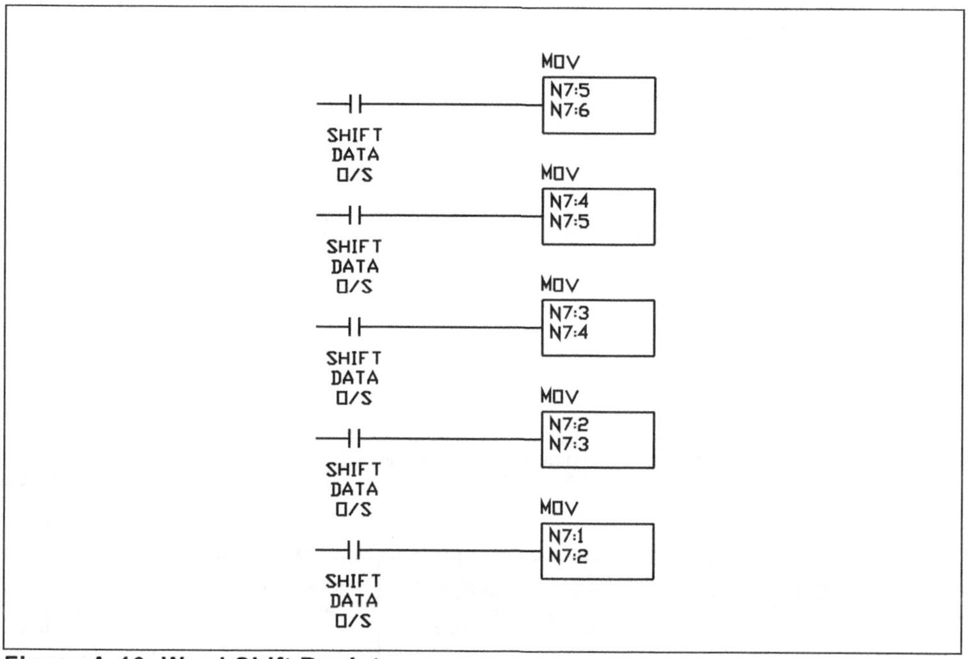

Figure A-10. Word Shift Register

Bit 0 - Over All Good / Bad
Bit 1 - Bolt Loaded / No Bolt
Bit 2 - Bolt Pressed / No Press
Bit 3 - Bolt Press Reached Depth / Insufficient Depth
Bit 4 - Bolt Press Force Achieved / Insufficient Force
Bit 5 - Keeper Ring Loaded / No Ring
Bit 6 - Keeper Ring In Position / Out of Position
Bit 7 - Spare
Bit 8 - Spare

Figure A-11. Word Shift Register Bit Representation

can refer to a file containing many words, using the indirect addressing capability of the processor's instruction set.

A.3 Capturing Motion Time

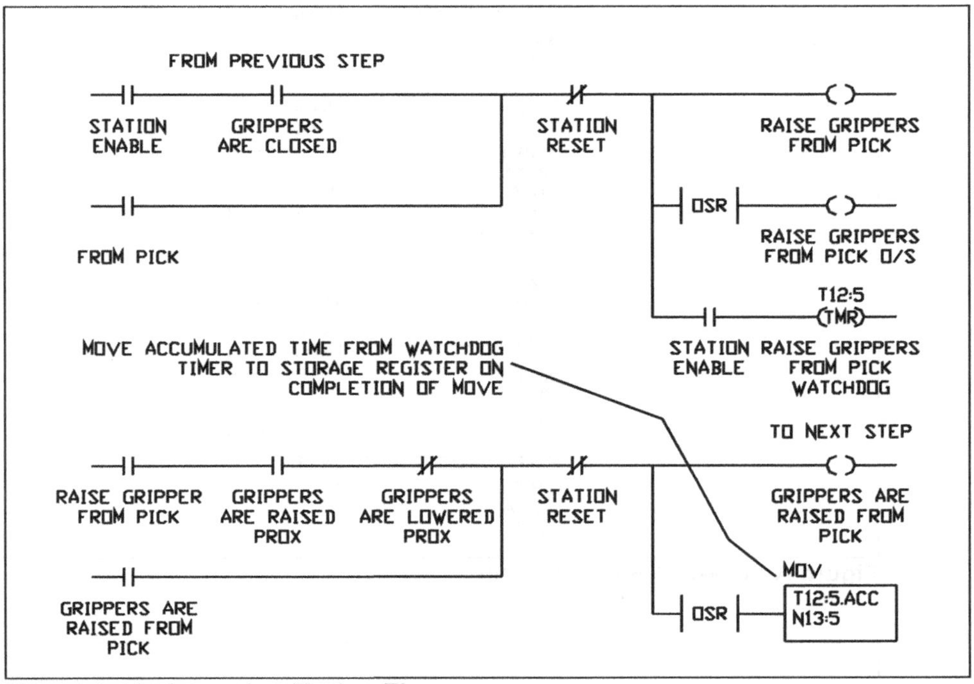

Figure A-12. Capturing Motion Times

A.4 Missed 3 in a Row Counter and Fault

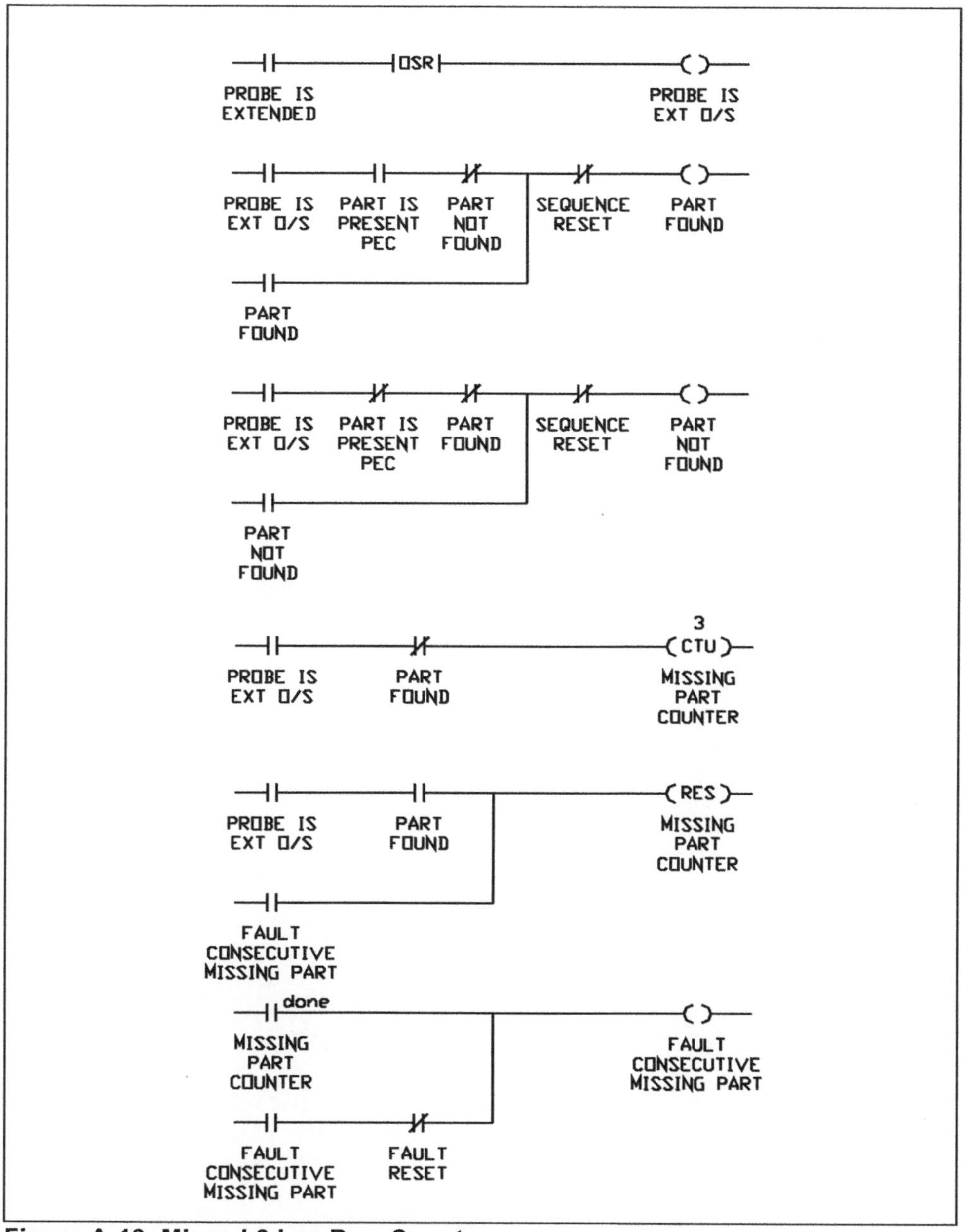

Figure A-13. Missed 3 in a Row Counter

Glossary

Air Dump Valve – A valve wired into the master circuit to relieve all pneumatic pressure from the machine when the master circuit is de-energized. The master circuit is de-energized when *Emergency Stop* push buttons are pressed and when guard doors are opened.

Ambush – An ambush occurs when an output latch is set, but the prime mover controlled by the output does not move. The prime mover may not move because the air supply is off via the lockout valve. When the air pressure is restored, which could be a long time later, the cylinder will then move. Since pressing the button to initiate a move and restoring air pressure are unrelated events, the cylinder can lay in wait, ready to act long after the operator forgets about it, creating a potential hazard.

Asynchronous – Asynchronous refers to independent operations, as the term is applied to equipment. Equipment with multiple stations is asynchronous if the stations are not timed to operate with each other. Each station operates as needed, depending only on the flow of product to them.

Auxiliary Equipment – Auxiliary equipment is the equipment used to support the processes of the machinery, such as dust collectors, vacuum pumps, hydraulic units, and so forth. It is usually turned on separately from the machinery.

Back Checking – Back checking is the process of looking at a sensor when the sensor is not needed, to check its operation. Say you have a proximity sensor that is extended to the top of a component part to see if a bolt is installed. If the sensor should fail, it may indicate a bolt is always present. The sensor can be checked when the component part is not present (thus no chance of a bolt being present) to make sure the sensor is de-energized.

Belt and Suspenders – Belt and suspenders refers to having redundancy in the logic, circuits, and devices on the equipment.

Big Table Theory – The big table theory is really a theory of preparation and organization. The equipment commissioning processes is an involved process that can make or break a project, especially with special, one-of-a-kind pieces of equipment, so be prepared. I like to set up a large table next to the equipment and place on the table the support PC, tools, and all the documents I may need. Useful documents include program printouts, electrical and pneumatic prints, timing diagrams, cut sheets, and product drawings.

Bit – A bit is a single element of a byte or a word memory unit, and it is either on (logically true) or off (logically false).

Board Game – The process of tracing through the sequence logic, using flow charts to see how the sequences interact with each other. The process is used to see how robust the sequences are and to spot possible interlock problems.

Branch (Rung) – A branch of a ladder logic rung is one set of logical conditions that is combined with another set of conditions to create a logical OR circuit. Logical conditions are typically normally open and normally closed contacts. Contacts in series form AND circuits, and branching in ladder logic forms OR circuits.

Branch (Sequence) – The flow of a sequence from the beginning to the end can have multiple possible routes or branches. The logic to determine which path to follow is called *decision logic.*

Bump-Less Transfer – The act of changing the operational mode from manual mode to automatic mode without the controlling device or equipment performing an action. It does not look very good to have a cylinder extending and retracting whenever the mode selector switch is used.

Byte – A byte is a memory unit consisting of eight bits.

Campaigning – Campaigning is the processes of building consensus to your ideals or it could be self-promotion. I really hate to see aluminum used for sensor flags and hard stops so I had to do a little campaigning in the mechanical department to persuade people not to use it.

Camping Out – Camping out is the processes of viewing the logic inside the controller as it executes, watching for an event causing the logic or equipment to misbehave. The use of fault latches to trap error conditions eliminates the need to camp out and watch the logic to determine what went wrong.

Child Sequence – A child sequence starts when commanded by a parent sequence and only operates once for each command from the parent sequence. Stations on a dial machine are controlled by a child sequence with the dial being the parent sequence. The dial sequence rotates the dial then tells each station to operate.

Clear Out Mode – Clear out mode is a special operating mode used to flush product out of a machine. Clear out mode tells the equipment not to accept any new product, finish processing the product already inside the machine, and cycle stop after the last piece is finished. Clear out is used when changing over to a different product or before shutting down for a holiday.

Coil – Coils are the components of an electromagnetic or solid-state relay used to change the state of the relay's contacts when energized or de-energized. A coil is also an instruction type of the programmable logic controller's instruction set. The functionality of the instruction mimics that of the relay's coil.

Collision – A collision is two or more motions that smashed into each other and is usually caused by failing sensors.

Commissioning – Commissioning is the process of making equipment work for the very first time. This process is also referred to as "startup," but many people call it "debug" because there are usually many problems. In theory, you should only need to set up and adjustment the equipment, instead of re-drilling holes and making program changes. In practice the theory should be our goal, so I prefer the term startup or commissioning instead of debug. The term debug implies original sin, and I like to think I was a little bit more prepared.

Component Parts – Component parts are the individual elements that make up the product that the machine is processing. If the equipment is assembling toner cartridges, then the component parts are all the cartridge housing pieces, the toner, labels, and so forth.

Contacts – Contacts are the components of an electromagnetic or solid-state relay that change states when the coil of the relay is energized or de-energized. A contact is also an instruction type of the programmable logic controller's instruction set. The functionality of the instruction mimics that of the relay's contact.

Cycle – A cycle is one iteration of the equipment's mechanisms required to perform a process. A cycle is characterized by the mechanisms returning to their starting positions, so the cycle can be repeated. When the equipment is operating in automatic mode, the cycle is continually repeated. A part loading station, for instance, picks a part and transfers it into the machine, then returns back home, so a new part can be picked on the next cycle.

Cycle Stop – A cycle stop brings the machine to rest at some predefined cycle stop point. It is used to stop the equipment in an orderly manner. Cycle stopping allows the equipment to finish the current machine cycle but not to start a new cycle. The most common cycle stop point is where the machine stops after the mechanisms return to their initial (home) positions.

Cylinder – A cylinder is a mechanical device that uses air pressure or hydraulic fluid to push a rod into or out of the body of the cylinder. The rod is connected to a mechanism and is used to move the mechanism between several positions.

Debug – The process of resolving problems. There are many methods and techniques, depending on the type of problem, and I suggest learning about them; they are helpful and will save you time.

Decision Fault – A decision fault is generated if none of the decision latches within a decision group turn on.

Decision Group – A decision group is two or more decision latches used to control branching within a sequence.

Decision Latch – A decision latch is a type of sequence latch used within a group of latches (decision group) to control what part of a sequence is executed. Only one of the decision latches within the decision group can turn on per cycle. The sequence branch associated with the winning decision latch is the one that is executed. The remaining sequence branches are prevented from executing. For example, decision latches are used to determine if a component part is either good or bad, remembers the decision, and controls if the part is used or rejected.

Decision Logic – Decision logic is made up of decision latches and a decision fault. The logic is used to control sequence branching.

Default Position – The default position is the position the output logic will place a prime mover when the processor is first powered up.

Dependent Sequence – A sequence that works with another sequence to perform a task. The two sequences are generally independent from each other, except when they need to interact. A vibratory feeder bowl with a part sorter and a pick/place mechanism that transfers parts to a station on a dial is a good example. The bowl, sorter, and pick/place mechanism operate to present a part, then it interacts with the station's mechanisms when the station asks for the part. The bowl, sorter, and pick/place mechanism is programmed, with its own sequence separate from the station's sequence to allow it to find and present a part while the station is busy or unable to operate, such as when the dial is indexing. Parent/child sequences are dependent sequences but are given their own classification due to the unique interlocking requirements between these types of sequences.

Detented – A detented valve uses two solenoids to position the spool inside the valve. The spool remains in its last position (detented) when power is removed from the solenoids. Emergency Stop push buttons are also detented, in that once pushed in they have to be pulled back out.

Dial – A dial is a mechanism that uses a circular plate to transfer product from one station to another. The product to be processed is located and held by fixtures or nests equally located around the perimeter of the dial. The stations are located around the dial and interact with the product in the fixtures after the dial stops rotating. The collection of dial and stations is referred to as a dial machine. The dial is usually coupled to a cam-style indexer that is driven by a motor, but direct drive servomotors are becoming more popular. The output of the indexer is attached to the dial and provides a dwell period where the dial is stationary, and an index period where the dial is rotating. A typical indexer is set up with a dwell period of 90° of input shaft rotation and an index period of 270° of input shaft rotation. A motor is connected to the input shaft of the indexer through a clutch/brake and a gearbox. The gearbox provides the mechanical advantage and the clutch/brake is used to start and stop the dial's rotation. A clutch/brake is used when the motor is constantly running and is eliminated if the motor is started and stopped, as in the case when a servomotor is used. The beauty of the indexer is its output shaft, and thus the dial does not rotate even when the input shaft continues to rotate during the dwell period. This provides a relative sloppy positioning device, such as a clutch/brake time to stop the input shaft and keep the dial from rotating when the station mechanisms are engaged. A sensor on the input shaft of the indexer senses the dial is in the dwell period and stops the motor or disengages the clutch/brake. The stopping device has the whole dwell period to bring the dial to rest. The indexer also provides consistent location of the fixtures on the dial. The fixtures end up in the same position, no matter where the indexer is during the dwell period.

Double-Ended Valve – A double-ended valve uses two solenoids to position the spool inside the valve. The spool remains in its last position when power is removed from the solenoids, except for three position valves where the spool is sent to the center position.

Downtime – Downtime is the time the equipment is not processing product when it should be. Downtime can be caused by unreliable mechanisms, inconsistent product, or worn-out mechanisms.

Dry Cycle – Dry cycle is a special operating mode where the equipment operates as it would in automatic mode but without product. Part hoppers and so forth are emptied to prevent components from being picked and placed into the machine and product sensors are bypassed. Dry cycle is useful for testing sequences and to wear test the equipment.

Dwell – The dwell portion of an indexer's cycle is when the output shaft does not rotate, even though the input shaft continues to.

Dwell Station – A dwell station operates while the dial on a dial machine is in the dwell portion of its rotation, as opposed to an indexing station that operates while the dial plate is rotating.

Event-Driven Logic – Each task of event-driven logic is signaled to start by the completion of the last task, as indicated by an event such as a proximity switch energizing. The opposite is time-driven logic where tasks are signaled to start based solely on time.

Fail Safe – A fail-safe circuit or component fails in such a way that the failing condition is considered safer than the other possible conditions the circuit or component could fail to. An example of a fail-safe component is a clutch/brake that will, on loss of power, engage the brake. The alternative is a clutch/brake that will fail with the clutch engaged, and as such, will not work when needed and will provide no warning of impending doom. An example of a fail-safe circuit is a normally closed proximity switch that de-energizes the input to the programmable logic controller when it detects the monitored condition. If the proximity switch's wiring should break or its power supply should fail, then the input is also de-energized to provide a warning.

Fault – A fault is an error condition that the logic detected and captured with a fault latch. Faults are problems with the equipment that require the equipment to stop. Faults are also referred to as alarms. Examples include overload relays tripping and watch dog timers expiring. The station generating the fault is stopped immediately, while the remainder of the equipment is cycle stopped.

Fault Latch – A fault latch is a latch that turns on (tripped or active) when a fault is detected. It turns off (reset or acknowledged) when the operator presses the fault reset push button, provided the faulting conditions are no longer present. The logic for a fault latch is configured such that the fault latch can only be reset when the conditions causing the fault are gone. A "low air pressure" fault can only be reset when the air pressure is once again at a proper pressure. This prevents personnel from restarting the equipment until the problem is corrected.

Fault Recovery – Fault recovery is the process of detecting errors and providing functionality to deal with the errors. Error detection includes monitoring each initiated action to determine if it finished or failed. Error handling includes providing tools for the operator to speed recovery such as the ability to abort and restart sequences, manual mode operations, stepping, and single cycling. Since the goal is to minimize downtime, error handling also includes providing convenience functions to the operator, such as the ability to home a station or the entire machine with a single push button.

File – A file is a group of bits or words sharing a common address structure. The address structure allows the individual bits within the file to be accessed individually, accessed in a group as a word or accessed all at once as a file.

First Power – First power occurs when the master circuit is energized after the controller is initialized. In this condition, the outputs may not be synchronized with the prime movers attached to them. The output logic defaults to position the prime movers one way when the controller is initialized but the prime movers may not be in this default position, thus causing unwanted motion when motive power is applied.

Fixture – A fixture is a component of the machinery used to hold and locate the product that the machine is working on. A fixture is also referred to as a nest. A typical fixture has features to hold the product in the correct orientation and locates the product to a known reference point. Many fixtures also include features to allow large forces to pass through to the product. When press fitting two parts together, a backup cylinder or anvil is located under the fixture to resist the insertion force.

Flag – A flag is a feature on a mechanism used to trigger a sensor to change states.

Flow Chart – A flow chart is a graphical representation of how the logic solves itself. It shows what conditions the logic waits for, what decision the logic makes, and what the logic accomplishes.

Forward Checking – Forward checking is a method of checking a sensor's functionality. The sensor is checked just before the sensor is needed. For instance, a "cylinder is extended" sensor is checked to make sure its de-energized before the cylinder is commanded to extend. If the sensor is on, then a fault is generated. Without the forward check, the cylinder may report it is extended immediately after being told to extend.

Freezing Cycle – Freezing cycle is to stop a sequence from executing its steps. A sequence that cannot execute is said to be frozen or holding cycle.

Frozen Sequence – A frozen sequence is a sequence that has started to execute its steps but is currently being prevented from continuing. A frozen sequence will not initiate any new actions but will continue to monitor its last action until that action either finishes or fails. Sequences are frozen when a problem with the sequence occurs and when stepping through a sequence.

HMI (Human–Machine Interface) – A programmable device that features a monitor for displaying information and some type of input device, such as a keyboard, keypad, or touch-sensitive overlay on the monitor. Typical use to replace hardwired push buttons, pilot lights. and readouts.

Holding Cycle – A holding cycle stops a sequence from executing its steps. A sequence that cannot execute is said to be holding cycle or frozen.

Home Position – The position a sequence expects a mechanism to be at when the sequence starts to execute. Checking to make sure the mechanisms are home before a sequence starts keeps the sequence and mechanisms synchronized. Damage to mechanism may result when a sequence command motions and the motions are not where the sequence thinks they are.

Homing – The process of positioning motions to their starting positions. Mechanisms can be commanded to move individually in manual mode or from homing routines included to speed up the process.

Hot and Online – Hot and on line refers to the support software running and connected to the programmable device on the equipment so the operation of the device can be seen. When debugging the communication between two programmable devices, such as a programmable logic controller and a motion controller, it is easier to be hot and online to both at the same time to see where signals fail to pass.

Independent Sequence – An independent sequence starts when the system control starts and repeats as long as the system control stays on. Independent sequences are used to control groups of mechanisms, which are independent from the other mechanisms on the machine. A vibratory feeder bowl with a part sorting mechanism that sends parts into a machine on a conveyor belt is a good example. The bowl and sorter operate to keep product at the end of the conveyor and does not really care about the mechanisms taking parts off the conveyor. The conveyor provides the buffering of parts and eliminates the need for the bowl and sorter to interact directly with the rest of the equipment.

Indexer – An indexer is a mechanical device consisting of gears and cams. It is used to provide intermittent motion on its output shaft with constant motion on its input shaft. An indexer is typically used to rotate a dial plate on a dial machine.

Index Station – An index station operates while the dial on a dial machine is in the index portion of its rotation, as opposed to a dwell station that operates while the dial plate is stationary. Index stations typically consist of noncontact sensors, such as photoelectric eyes looking for errant product in fixtures or splices in webs.

Inspection – An inspection is a process that examines the component part that the equipment is processing to see if the product meets acceptable criteria.

Interlock – Interlocks are a set of conditions that must be met before an operation can take place. The manual mode layer includes interlocks to prevent the operator from initiating a motion if that motion will crash into another mechanism.

Indirect Addressing – Indirect addressing is the ability to reference a bit, word, or an entire file based on the value in another word. The logic leading up to a coil either turns the bit addressed by the coil on or off when indirect addressing is not used. With indirect addressing, the coil's address contains a reference to another word where the value addressed by the word tells the processor which bit to turn on or off. The value of the word can change, so the next time the rung is evaluated a difference bit can be changed.

Jams – The product or the product in combination with a motion is bound up and stuck. This can be a dangerous situation because of the motion still being powered, such as when a part is wedged in front of an air cylinder exerting its full force. The part can suddenly become free with the full force of the motion now snapping at your fingers.

Ladder Logic – A programming language used in programmable logic controllers. The language mimics control circuits made up of contacts, coils, and timers.

Latch – A latch is either a type of ladder logic output coil instruction or a whole rung of ladder logic. As a rung of ladder logic, it is an output instruction that turns on when specific conditions are met (setting conditions) and stays on until specifically told to turn off (reset). The latch rung stays on with the help of a holding branch. The three general types of latches used in Cascading Logic are the sequence latch, the fault latch, and the decision latch. As a type of output coil, the latch represents a memory element that turns on and stays on until instructed to turn off with an unlatch output coil. Latch and unlatch output coil instructions have many different names, depending on the type of processor used such as Set/Reset and L/U.

LVDT – LVDT is an abbreviation for a linear voltage differential transformer. It is an electrical device used to report position. The output of an LVDT is an analog signal that is proportional to the distance the plunger of the LVDT is moved from its zero point.

Machine States – Machine states are the operational conditions of the equipment, such as running, manual mode, or faulted. They are used to consider how the equipment responds to changing conditions and how the equipment is navigated between the states.

Manual Mode or Manual Mode Layer – Manual mode functions control each prime mover on the equipment individually, such as extending or retracting a cylinder or jogging a motor. Manual mode functions are interlocked to prevent an action from initiating when that motion will interfere with another mechanism, when the machine is in the wrong operating mode, or when operator safety is a concern. The logic associated with manual mode operations is typically called the manual mode layer and is independent of the sequencing logic.

Master Control Circuit and Master Control Relay (MCR) – The master circuit is a relay or group of relays used to control all motive power. Power to all the motions is removed when this circuit is off. By removing power, a failure in a servo drive, for example, cannot cause the motor to run away. An operator is still required to perform lockout and tagout, but dropping the master circuit whenever an operator enters the machine provides additional safety.

Mechanism – A mechanical component on a machine, or a group of mechanical components. Examples of individual components include gears, linkages, ball screws, slides, and cylinders. An example of a group of mechanisms is a pick/place unit.

Message Bits – Message bits are any coils that are required to display a message to the operator whenever it is on. Typical message bits include fault latches from the motion objects and other coils used to indicate machine status, such as machine running or low product hoppers. The machine program scans all the message bits and sends a corresponding message to a display unit for each active message bit.

Motion – The combination of a prime mover and a mechanism that causes movement. A pneumatic cylinder attached to part press is a good example of a motion. The word "motion" generally describes devices that could cause injury because something usually has to move to cause harm, but heat sources can also be considered a motion.

Motive Power – Motive power is the energy source a prime mover needs to operate. Electricity, air pressure, and hydraulic pressure are common sources of motive power. Motive power is removed from the equipment when the master circuit is de-energized.

N.C. – Normally Closed

Nest – A nest is a component of the machinery used to hold and locate the product the machine is working on. A nest is also referred to as a fixture. A typical nest has features to hold the product in the correct orientation and locates the product to a known reference point. Many nests also include features to allow large forces to pass through to the product. When press fitting two parts together, a backup cylinder or anvil is located under the nest to resist the insertion force.

New Operator Syndrome – New operator syndrome refers to how the equipment operates differently when a new group of people operates it. Problems with the equipment sometimes show up after the equipment is installed at the end users facility because the operators run the equipment differently than you did while you were commissioning it. The operators may press push buttons in a different order or may not be aware of critical steps.

N.O. – Normally Open

Nonrecoverable Fault – A nonrecoverable fault is a problem with the equipment that cannot be easily solved and usually requires the equipment operations to be aborted. Examples included a busted mechanism and worn-out air cylinders.

Operator Interface – The collection of push buttons, pilot lights, human–machine interfaces (HMI), and so forth that allow personnel to operate and control the equipment.

O/S – O/S stands for one shot. It is a ladder logic instruction used to provide a pulse when a signal first turns on or off. The duration of the pulse is only one program scan long. A rising or leading one shot occurs when the signal first turns on, and a falling or trailing one shot occurs when the signal first turns off.

Pallet – A pallet is basically a movable fixture. The pallet transports the product from one station to the next and can include methods to track and store information about the product, such as bar code labels, radio frequency chips, or mechanical pass/fail pins. Pallets are usually used on asynchronous machines and ride on belt conveyors. Mechanical stops are used to stop the pallet at a station or to let it pass untouched. The conveyor can move the pallet, but if you hold the pallet, it will slip on the conveyor and remains stationary, hence the phrase "power and free" to refer to this type of system.

Parent Sequence – A parent sequence controls other (child) sequences. Parent sequences control the mechanisms in common to all the other groups of mechanisms on the machine. They start when the system control starts and repeat as long as the system control stays on. The sequences controlling a dial or a walking beam are good examples. When these mechanisms finish transferring parts between the stations, they then command the stations to operate.

Part Tracking – Part tracking is the logic used to maintain the information associated with the product the equipment is processing. The information is advanced with the product as the product travels through the equipment with a shift register. The information may be a single bit used to indicate if the product is good or bad. It can also be a collection of words and bits that captures data for archiving, such as insertion forces, sealing temperatures, and so forth.

PEC – PEC is an abbreviation for a photoelectric sensor.

Photoelectric Sensors – A photoelectric sensor uses light to detect the presence or absence of an object. Photoelectric devices come in all shapes and sizes. It is best to work with a manufacture's representative when picking these types of sensors. Various sensors can be tested with the actual component parts to pick the optimal sensor for the application. Remember that the reliability and functionality of the whole machine can boil down to one sensor, so take the time up front to pick the correct sensor.

Pick/Place Mechanism – A pick/place unit is a group of mechanisms used to pick a component part from one location and place it at another location. Typical pick/place mechanisms have three motions: a gripper to grab and release the component part, a vertical motion to raise the part above obstacles such as the machine frame, and a horizontal motion to relocate the part. Vacuum cups are also used instead of a gripper to hold the component part.

PID – PID stands for proportional, integral, and derivative. It represents an algorithm used to control temperatures and servomotors.

Position Bit – A position bit is a rung of logic that describes the position a mechanism is currently at. It is the summation of all the sensors associated with the mechanism, logically combined to reflect the state of each sensor for the particular position the position bit is describing. It is used to enhance "back checking" and "forward checking" of the sensors. The position bit will remain off if any sensor should fail to provide a means for the logic to detect a failed sensor.

Power and Free System – A power and free system is a mechanical system where conveyor belts or chains transport the product through the equipment. The product is usually held in some type of carrier or pallet. The driving mechanism moves the product, unless it is held with simple stop gates or similar mechanism.

Power Down Reset – A power down reset is the act of turning the main disconnect off, or turning power to the processor off, to recover from a problem. Poorly programmed equipment may stop and fail to respond to operator action, leaving the operator and maintenance personnel no other choice than starting over completely from scratch.

Prime Mover – A prime mover is any device that can transfer energy, such as motors, pneumatic and hydraulic cylinders, heaters, and ultrasonic welders.

Product – The product is the materials that the equipment is processing, such as cup cakes or camera film.

Programmable Logic Controller – A programmable logic controller is a programmable device that can read inputs to gather information, execute a program, then set outputs to control devices. The typical programming language is ladder logic.

PRS – PRS is an abbreviation for a proximity sensor.

Proximity Sensor – A proximity sensor or switch is a device that uses an electric or magnetic field to detect the presence of an object. A magnetic field is used in inductive proximity switches to detect metal objects, and an electric field is used in capacitive switches to detect nonmetallic objects such as plastic or wood. The output of the sensor changes states when an object (flag) is in the switch's sensing range. A common use of a proximity switch is to provide positional information of a pneumatic cylinder. One switch is used to detect when the cylinder is extended and another is used to detect when the cylinder is retracted.

Purge Mode – Purge mode is a special operating mode used to flush product out of a machine. Purge mode tells the equipment to not accept any new product, finish processing the product already inside the machine, and cycle stop after the last piece is finished. Purge is used when changing over to a different product or before shutting down for a holiday.

Push Button – A push button is an operator interface device used to provide or remove a signal when pressed. It typically refers to a discrete device mounted on an operator control station or it can be a touch-sensitive cell on a HMI.

Push Pull Logic – Push pull logic refers to output logic configured to prevent both solenoids of a double-ended valve from energizing at the same time.

Recoverable Fault – A recoverable fault is a problem with the equipment that can easily be fixed to allow the equipment to resume operations. Examples included realigning product and readjusting sensors.

Resuming Cycle – Resuming cycle is to let a frozen sequence, or the whole machine, continue to execute and process the product.

Retentive Memory – Retentive memory are memory elements that maintain there on state, off state, or value when the power to the programmable logic controller is cycled, or when the mode is changed. Retentive memory is typically maintained with the programmable logic controller's battery and is sometimes called battery-backed ram.

Rung – A rung is the smallest division of a ladder logic program. The accumulation of logic in a rung leads to some type of output instruction such as a coil.

Scan Time – Scan time is the cyclic tasks performed by a programmable logic controller. A typical scan consist of reading inputs, executing logic, updating outputs, communications, and housekeeping functions. The scan time as expressed in specifications is only the time it takes to execute the logic and does not include the time required to perform the other activities. It is usually expressed in ms/k (milliseconds per one thousand words) of executed logic.

Sequence – A sequence is an ordered list of actions or tasks the machine is required to perform, where each action or task is a step of the sequence. When dealing with a group of prime movers and mechanisms on the machine, a sequence tells them exactly what to do. Sequences are usually cyclic, in that the controlled mechanisms end up where they started from, so the sequence can be executed over and over again. Sequences are typically illustrated with flow charts, described with text as a theory of operation or as a control functional description. Sequences can be categorized based on who controls them and how they interact with other sequences, as either a parent, child, independent, and dependent sequence.

Sequence Fault – A sequence fault is a fault latch that is set when conditions required for a sequence to proceed fail to occur within a time limit. The sequence fault is used to prevent a sequence from stalling by providing a path the sequence can follow when it cannot follow its normal execution path. The sequence fault also alerts the operator of the failure. The sensor indicating a cylinder is extended needs to signal that the cylinder extended after it is commanded to extend. Otherwise, the sequence will wait forever for it without the fault latch.

Sequence Latch – A sequence latch is used in the construction of sequences to initiate actions, remember what actions were initiated, and to remember what actions successfully finished. The primary attribute of a sequence latch is the unconditional resetting of the latch when told to do so, unlike a fault latch that can only reset when the conditions that set the latch are no longer present.

Servomotor – The term servomotor used in this book refers to the motor, drive, and motion controller comprising a complete servo subsystem.

Single Cycle – Executing a sequence just once on demand from the operator interface. Sequences typically execute over and over, as the equipment operates in automatic mode but can be programmed to operate just once. Single cycling is useful during startup and debug to check individual stations on a machine, to see how all the mechanism interact at rated speed. Single cycling is also used to command a frozen sequence to restart and complete.

Solenoid Valve – A solenoid valve is an electrical/mechanical device used to control the flow of air or hydraulic fluid to a cylinder. Electrical signals control one or two solenoids that shift the position of an internal spool. The spool redirects the flow of air or fluid between the ports on the valve.

Spring Return Valve – A spring return valve uses a spring to return the valve's spool to a default position. A solenoid or piston is used to move the spool against the spring, reconnecting the ports of the valve.

Spool – A spool is an internal component of a valve used to redirect the flow of air or hydraulic fluid between the ports on the valve.

Spool Position – The spool position is the position of the spool inside the valve and is used to express the position or action of the attached prime move. The spool of a double-ended valve will remain in its last position when both solenoids of the valve are de-energized, giving the valve a memory element. It is important to know the spool position when the signals to the solenoids are interrupted, say when the master circuit is de-energized and re-energized, to predict if the spool will shift and thus create motion on the equipment.

Spooling – Spooling is the act of using an object like a screwdriver to reposition the spool of a valve to force the valve to operate. Many valves expose the ends of the spool to allow this, and some have manual operators to make this process easier to perform.

Stalled Sequence – A sequence that can no longer proceed in executing its steps and provides no indication that it stopped. A fault latch in combination with a watch dog timer is usually used to indicate when a set of conditions inhibits a sequence from proceeding. The fault latch provides a path for the sequence to continue, either by resetting the sequence directly or by alerting the operator to make a choice.

Station – A station is a subsection of a machine that typically performs a single operation on the product. The logic for a station is basically independent from the logic for the remaining equipment, except for some part transfer handshaking, interlocking, and data transfer.

Step – A step is a single action or task of a sequence. When dealing with actions, a step is the smallest action the machine controller can initiate, such as energizing a solenoid valve or triggering a vision system to start its inspection routine. When dealing with tasks, a step could decide if an inspection passed or failed, or it could be saving data to a database.

Stepping – Stepping is the process of executing a sequence one step at a time. This allows checking of the sequence's programming, and the alignment of the mechanisms working with the product.

Summation – Summations are useful to collect information and use it elsewhere in the logic as a single bit. Examples include the System Fault Summation and a Station Home Summation. A summation can be an OR circuit, where the output of the rung is on whenever any bit comprising the circuit is on. They can also be an AND circuit, where the output of the rung is on when all the bits that make up the circuit are on.

Support Software – The software used to support the programmable devices on the equipment, such as the programmable logic controller, HMI, and motion controllers. Programmable logic controllers have companion software, used to create the program for the controller, which allows transferring of the program to and from the controller, viewing the logic in the controller as it executes, and can even allow changes to the logic as the controller executes.

SVSR – An abbreviation for a solenoid valve spring return.

SVDD – An abbreviation for a solenoid valve double-ended, detented.

Synchronous – Synchronous refers to dependent operations as the term is applied to industrial machinery. Equipment with multiple stations is synchronous if the stations share timing or their actions are otherwise coordinated. A good example of a synchronous machine is one in which all the motions are driven from a common line shaft.

System Control – The system control refers to the logic used to control the machine as a whole in automatic mode and the logic common to all the individual sections of the machine, such as the safety circuits.

Thawing Cycle – Thawing cycle is to let a frozen sequence or the whole machine to continue to execute and process the product.

Throughput Time – The throughput time is a measure of how fast a programmable logic controller can respond to an event. It is the time between when a sensor is triggered to when a physical output of the controller changes state. It includes the response times of the sensor, input module, controller (including its scan time), and the output module. The throughput time is used to check if events occur faster than the controller can detect or control them.

Time-Driven Logic – The tasks of time-driven logic are signaled to start based solely on time. The opposite is event-driven logic where each task is signaled to start by the completion of the last task as indicated by an event such as a proximity switch energizing.

Walking Beam – A walking beam is a product transfer mechanism used to move product from one station to another. The walking beam either moves the product directly, or moves the fixture or pallet holding the product. Walking beams typically move the product between stationary nests using two motions: one motion to lift the product and the second motion to move it horizontally.

Watch dog Timer – A watch dog timer (Dog for short) is used to monitor an event to determine if the event failed. The event has only so much time to report it finished, or the watch dog timer expires and triggers a fault. Watch dog timers can be used to monitor single events, such as a cylinder extending, or a group of events, such as a whole sequence.

Word – A word is a memory unit consisting of two bytes or 16 bits.

Index

Printed and bound by CPI Group (UK) Ltd, Croydon, CR0 4YY
15/09/2026
14962414-0001